THE PRACTICAL GUIDEBOOK

THE PRACTICAL GUIDEBOOK
OF
ESSENTIAL ISLAMIC SCIENCES

A COMMENTARY ON IBN ASHIR'S
AL-MURSHID AL-MU'IN

by Shaykh Ali Laraki al-Husaini

MEEM INSTITUTE

The Practical Guidebook of Essential Islamic Sciences

Published by: The Meem Institute

By: Shaykh Ali Laraki
Edited by: Hajj Issa Bryce
Typeset by: Abdassamad Clarke

A catalogue record of this book is available from the British Library.

ISBN-13: 978-0-9574435-0-1 (Paperback)
978-0-9574435-1-8 (e-book)

Printed by Publidisa

DEDICATION

To my shaykh, Shaykh Dr. Abdalqadir as-Sufi

Acknowledgements

I want to acknowledge the colossal effort undertaken by my student Yakub Werdelin and his patience throughout the process that culminated in the completion of this book. May Allah reward him for it. Amin.

I am also deeply grateful to: Hajj Issa Bryce and Yusuf Rowland for their altruistic proofreading of the text, my daughter Naila for her invaluable help, my student Mehmet Currie, to whom I am especially indebted for his generous help and assistance; there is hardly a page which has not profited by his assistance, and Mr. Syed Anwar al-Banna and Mr. Zulfiqar Awan for making financially possible the publishing of this book.

May Allah reward all of them. Amin.

Contents

بسم الله الرحمن الرحيم

والصلاة والسلام على سيدنا محمد

وعلى آله وصحبه أجمعين

مَنْ يُرِدِ اللَّهُ بِهِ خَيْرًا

يُفَقِّهْهُ فِي الدِّينِ

Whomever Allah wills good, He grants him deep understanding (fiqh) of the Deen.

(Malik, Muslim, al-Bukhari)

Preface by
Shaykh 'Abd al-Hayy al-'Amrawi

Translated from the Arabic

In the name of Allah,
the Merciful and Compassionate
Blessings on the Messenger of Allah

CERTAINLY, MANY MUSLIM brothers and sisters, who do not know Arabic are in dire need of beneficial books like the one authored by 'Abd al-Wahid ibn 'Ashir al-Andalusi and its commentary, by Muhammad ibn Ahmad from Fez by origin, residence and upbringing, known as Mayyara, who based themselves on the three concepts that appear in the hadith of the Messenger of Allah ﷺ narrated by Muslim in his *Sahih*, from 'Umar ibn al-Khattab, who said:

"One day while we were sitting with the Messenger of Allah ﷺ a man came up to us whose clothes were extremely white, whose hair was extremely black, upon whom traces of travelling could not be seen, and whom none of us knew, until he sat down close to the Prophet ﷺ so that he rested his knees upon his knees and placed his two hands upon his thighs and said, 'Muhammad, tell me about Islam.'

"The Messenger of Allah ﷺ said, 'Islam is that you witness that there is no god but Allah and that Muhammad is the Messenger of Allah, and establish the prayer, and give the zakat, and fast Ramadan, and perform the Hajj of the House if you are able to take a way to it.' He said, 'You have told the truth,' and we were amazed at him asking him and [then] telling him that he told the truth.

"He said, 'Tell me about Iman.' He said, 'That you believe in Allah, His angels, His books, His messengers, and the Last Day, and that you believe in the Decree, the good of it and the bad of it.' He said, 'You have told the truth.'

"He said, 'Tell me about ihsan.' He said, 'That you worship Allah as if you see Him, for if you don't see Him, then truly He sees you.'

"He said, 'Tell me about the Hour.' He said, 'The one asked about it knows no more than the one asking.'

"He said, 'Then tell me about its signs.' He said, 'That the female slave should give birth to her mistress, and you see poor, naked, barefoot shepherds of sheep and goats competing in raising buildings.'

"He went away, and I remained some time. Then he said, 'Umar, do you know who the questioner was?' I said, 'Allah and His Messenger know best.' He said, 'It was Jibril who came to you to teach you your Deen'."

The Deen consists of three elements, as reflected in the aforementioned hadith, when the Messenger of Allah ﷺ said: 'It was Jibril who came to you to teach you your Deen'. These three elements are the ones contained in the poem of Ibn 'Ashir and its commentary by Mayyara. The first third is Iman, the second third is Islam and the third third is Ihsan, which is the adoption by the servant the manners that befit servanthood. And this is the science of Tasawwuf.

Our brother 'Ali Laraki has done a good thing translating this book into English in order to transmit to the people the rulings of their Deen. It will be a good thing if others follow his example and translate it into Spanish, French, German and Italian etc., so the speakers of those languages could also benefit by it.

The Muslim has to believe in the Messenger Musa – blessings and peace upon him – and the book revealed to him, the Torah, and to believe in the Messenger 'Isa – blessings and peace upon him – and the book revealed to him, the Injil, as he has to believe in all the noble Messengers – blessings and peace upon them. In Sura Al-Baqara, verse 284, it is stated: "*The Messenger believes in what has been sent down to him from his Lord, and so do the believers. Each one believes in Allah and His angels and His Books and His Messengers. We do not differentiate between any of His Messengers.*" This matter is what is contained in the first third of this book.

Our brother 'Ali Laraki, has responded to the saying of the Messenger ﷺ: "Transmit from me even if it is only one verse." He has, indeed, transmitted a number of verses, the first being the words of Allah – exalted is He – in Surat al-Ma'ida: "*O you who believe! When you get up to do salat, wash your faces and your hands and your arms to the elbows, and wipe over your heads, and [wash] your feet to the ankles. If you are in a state of major impurity, then purify ourselves.*" And has transmitted the words of Allah – exalted is He – in Surat al-Baqara 2:42: "*Establish salat and pay zakat and bow with those who bow.*" And he has transmitted the verse of fasting, stated in Surat al-Baqara 2:182: "*O you who have Iman! Fasting is prescribed for you.*" And he has transmitted the verse of Hajj, stated in Sura Al 'Imran 3:97: "*Hajj to the House is a duty owed to Allah by all mankind – those who can find a way to do it.*"

Despite not knowing the English language into which this book has been translated, I am very pleased by the matter. Therefore, I am very grateful to our brother 'Ali for this great work and I ask Allah to grant him success.

Shaykh 'Abd al-Hayy al-'Amrawi
Teacher at the Qarawiyyin Mosque, Fes

Preface

THE AUTHOR OF AL-MURSHID AL-MU'IN

HE IS THE great imam, the sea of knowledge, the mujahid, Abu Muhammad, 'Abd al-Wahid ibn 'Ali ibn 'Ashir, of the Ansari line, originally from Andalusia but brought up in Fes which he made his home.

He had vast knowledge of various disciplines, like Qur'anic readings (*qira'at*), Qur'anic writing (*rasm wa dabt*), grammar (*nahw*), morphology (*sarf*), syntax (*i'rab*), Qur'anic commentary (*tafsir*), hadith, theology (*kalam*), law (*fiqh*), tasawwuf, logic (*mantiq*), eloquence (*bayan*), poetic meters (*'arud*), time calculation (*tawqit*), arithmetic (*hisab*), principles of jurisprudence (*usul al-fiqh*), inheritance law (*fara'id*), medicine (*tibb*), and others.

He performed the pilgrimage to Makka, participated in jihad military campaigns, frequently retreated to the mosque for *i'tikaf* and performed *tahajjud* salats at night.

He has compositions on various knowledges, amongst them this incomparable poem that we have translated and commented upon, entitled *al-Murshid al-Mu'in 'ala ad-Daruri min 'Ulum ad-Deen* on Ash'ari 'aqida, Maliki fiqh and Junaydi[1] tasawwuf. In it he gathered both the roots and branches of the Deen to such an extent that whoever reads it and understands its points, is able to abandon once and for all the taqleed – imitation – of someone whose soundness of Iman is disputed, being able to learn the sciences that Allah has made incumbent on him.

He died in the city of Fes, where he is buried, on the 3rd of Dhu'l-Hijja 1040 AH / 3rd of July 1631 CE. May Allah cover him with His Mercy and place him in His spacious Garden.[2] Amin.

---

1 The biographies of Imams al-Ash'ari, Malik and Junayd will follow on pps. 2-3.

2 From *ad-Durr ath-Thamin*.

THE WORK

The work is the very famous poem entitled *al-Murshid al-Mu'in 'ala ad-Daruri min 'Ulum ad-Deen*. Its meaning is *The Helpful Guide to the Necessary Sciences of the Deen.*[3] This poem is memorised and studied in Morocco, where the author lived, and throughout North and West Africa. It is considered the first book to be studied on Islamic teaching in the *Madrasas* (Qur'anic Schools) after, or along with, the memorisation of the Noble Qur'an. It is an introductory treatise about Ash'ari belief, Maliki jurisprudence and Junaydi tasawwuf. It consists of everything a believer needs to know to acquire a basic knowledge of the Deen.

Initially the book was written to deal with the pillar of hajj, but the author introduced the rest of the five pillars of Islam with the addition of a chapter on tasawwuf. It was conceived as a poem to be easily memorised and for the illiterate.

The structure of the work is as follows:

1. General introduction presenting the author and the sciences to be mentioned (verses 1 to 5)
2. Introduction to the chapter of beliefs and a brief description of what a rational statement is and the signs of responsibility (*taklif*) (verses 6 to 13)
3. Exposition of the Muslim creed according to the school of al-Ash'ari (verses 14 to 47)
4. Introduction to the science of fiqh with definition of legal judgement and its ratings (verses 48 to 53)
5. Chapter on Purification (*tahara*) (verses 54 to 98)
6. Chapter on Salat (verses 99 to 181)

---

3 This word is normally translated as 'religion'. The reason why we are not going to translate it is because the word religion has the Judeo-Christian connotation of belief and relation with the Divine and their institutionalisation that we are trying to avoid. The Arabic word Deen has nuances of obedience, liability, subjection and indebtedness; in other words, the way a person transacts throughout his life, on the basis of what he believes reality to be. This will include all transactions between human beings themselves and the Divine. The word Deen also includes anything a person believes about this world and the next, regardless of whether or not they believe in God and the afterlife. Therefore, atheism, ecumenism and democracy are also Deens. Consequently, when we say that Islam is a Deen, we mean a religion, as well as all the social and political consequences derived from its worldview.

7. Chapter on Zakat (verses 182 to 210)
8. Chapter on Fasting (*sawm*) (verses 211 to 228)
9. Chapter on Pilgrimage (hajj) (verses 229 to 290)
10. Chapter on Tasawwuf (verses 291 to 312)
11. Epilogue (verses 313 to 317)

These eleven points are divided into three basic parts:

- Ash'ari 'aqida (creed)
- Maliki fiqh (jurisprudence)
- Junaydi tasawwuf

The reason why the work is divided into three parts is because they are the sciences ('aqida, fiqh and tasawwuf) that correspond to the three parts of the Deen mentioned in the famous hadith of Jibril. In this hadith – also called *Umm al-Ahadith* (the matrix of all hadith) – the Angel Jibril – appears physically to the Companions of the Prophet ﷺ, hence its importance– asks the Prophet ﷺ what is Islam, what is Iman and what is Ihsan. The Prophet ﷺ answers and upon the departure of Sayyiduna Jibril, he informs the Companions – who until that moment did not recognise the questioner – that it was Jibril who came "to teach you your Deen."

From these last words of the hadith, the 'ulama have concluded that the Deen is structured in three parts:

- Islam
- Iman
- Ihsan

Islam being what regulates the seven limbs that Allah has made us responsible for, namely:

- ears
- eyes
- tongue
- belly
- genitals
- hands
- feet

Iman being what our hearts must believe in.

And Ihsan or excellence of action and state achieved by travelling the spiritual path that starts with our awareness of being seen by the One Who sees everything (*muraqaba*), and ends with His contemplation (*mushahada*).

Therefore, the three sciences that correspond to the three parts of the Deen are:

- The science of 'aqida for Iman
- The science of fiqh for Islam
- The science of tasawwuf for Ihsan

This has been the traditional formulation of Islam throughout the whole of Muslim history, geography and the great centres of knowledge, like the universities of al-Qayrawiyyin, az-Zaytuna and al-Azhar, until innovators appeared in the last two centuries redefining Islam and misguiding people.

THE COMMENTARY

I thought initially to translate the text of the *Murshid al-Mu'in*, but very soon I realised that a commentary and an explanation of the verses would be more help to the average Muslim who is interested in consulting a reliable guide to put the Deen into practice rather than the linguistic intricacies and details of the translation of the text of the *Murshid al-Mu'in*. So, I provided a commentary of the *Murshid al-Mu'in* based a great deal on its most widespread commentary, the *Mukhtasar ad-Durr ath-Thamin* by Mayyara, and I tried my best to give it the structure of a modern manual that is visually and conceptually, easily accessible.

However, I have put in the margins of the commentary the number of the verse of the *Murshid al-Mu'in* that is the basis of the commentary, so those interested in the Arabic text can refer to it by going to the end of the book where they can find the full numbered text of the poem.

In my commentary, I have relied on the edition of the *Mukhtasar ad-Durr ath-Thamin* by Mayyara – also known as *Mayyara as-Sugra* – published by the Ministry of Awqaf and Islamic Affairs of Morocco as a primary source. I have also relied on its gloss (*hashiya*) by Ibn Hamdun ibn al-Hajj as a second reference along with the commentary *al-Habl al-Matin* by al-Marrakushi al-Muwaqqit. I have made wide use of the *Risala* by Ibn Abi Zaid al-Qayrawani, the *Aqrab al-Masalik* and the *Mukhtasar* by Khalil ibn Ishaq with their commentaries. Sporadic reference has been made to modern authors like al-Habib ben Tahir[4] and al-Ghiryani.[5]

REASONS TO UNDERTAKE THIS EFFORT

The commentary of the *Murshid al-Mu'in* is needed for English-speaking Muslims coming from the West, owing to the lack of knowledge of the Deen among Muslims in general and the damage done by the spread of deviant versions of Islam that increase confusion, alienation, division and hostility.

In sha'Allah, this work will provide an understanding of the Deen that will configure the essential structure needed by the student to appropriately categorise and correctly understand any further studies of the Islamic sciences.

Properly understood, this work can be considered a commentary on the previously mentioned hadith of Jibril. In it, the author explains and elaborates on the meanings of Islam, Iman and Ihsan, teaching what *every* Muslim should know about them.

As far as Iman and its corresponding science – 'aqida – is concerned, the poem will teach the correct way to believe in Allah, His angels, His books, His Messengers, the Last Day and the Decree, neutralising the false and anthropomorphic deviations that have been introduced among poorly informed Muslims, endangering their belief in Allah and impeding them from reaching true knowledge of their object of worship.

As far as Islam and its corresponding science – fiqh – is concerned, it teaches the way to learn the rules of 'ibada (worshiping) like the salat, zakat, fasting and pilgrimage (hajj) in the only rational way possible for a person who does not have any previous knowledge of fiqh; by adhering to and following one of the four acceptable schools of fiqh in operation today.

Not to adhere to a madhhab while learning fiqh has proven tantamount

4 *Al-Fiqh al-Maliki wa Adillatuh.*
5 *Mudawwanat al-Fiqh al-Maliki wa Adillatuh.*

to not learning any fiqh at all due to the confusion created in the mind of the student, who is incapable of understanding and apprehending legal rules without this necessary foundation.

In fact, not to adhere to a school is to reject the science of fiqh while claiming a superior understanding of the Deen to those great men who founded the madhhabs as well as their students who continued by expanding and fine-tuning their teachers' work.

The truth is that the followers of the founders produced the legal framework of the most glorious civilisations humanity has ever witnessed, and the non-madhhabi muftis have, in some instances, been the cause of confusion, social ailment and destruction.

What is important for every Muslim to understand is that a person who does not fulfil the preconditions of ijtihad is categorically obliged to adhere to a madhhab; the rigorous preconditions of ijtihad being as follows:

- To be Muslim
- To be responsible
- To have moral integrity (*'adala*)
- To be aware of the social circumstances of the time and place where we live
- To be perspicacious
- To master Arabic broadly. This implies grammar, lexicology and Arab rhetoric
- To know the Qur'an. This implies knowing the different causes of the revelation of the verses related to the law and whether abrogating or abrogated; to know each and every verse dealing with legal matters (500 verses approximately); to know the commentaries of the Prophet ﷺ on those verses and those made by the Sahaba, the Tabi'un and other scholars in the science of Qur'anic exegesis; to know the different readings (*qira'at*) of the Qur'an and the legal implications that their differences may originate
- To know the Sunna. This implies knowing all of the hadith which relate to the law, to know the abrogating and abrogated; to know the hadith and their transmitters; and the hadith that

specify and restrict other hadith in their legal implementation; to know the science of the different levels of hadith and their respective strengths

- To know the legal opinions of the Sahaba and the verification of their transmissions
- To identify the points of law where there is consensus or not
- To know the rules of analogy (*qiyas*) and its correct application
- To master the science of the Fundamentals of Fiqh (*Usul al-Fiqh*)
- To know the objectives of the Shari'a (*Maqasid ash-Shari'a*) and its application in order to balance the letter and spirit of the law and avoid rigidity that can lead to absurdity, unnecessary difficulty and injustice

This being the case, it is clear that the great majority of the Muslims do not fulfil these preconditions and are obliged to make *taqlid*, that is, to follow a mujtahid or his school and therefore, it is necessary to understand that *taqlid* can only be made of one of the recognised four schools of fiqh, and to make *taqlid* of other than them is forbidden by consensus.[6]

A layman cannot request legal evidence (*dalil*) regarding points of fiqh as if he had the capacity to understand it and judge its appropriateness. He must adhere to a madhhab and follow its Imam since he is the *dalil*.

As far as Ihsan and its corresponding science – tasawwuf – is concerned, its denial means abandoning a third of the Deen and its fundamental core.

To summarise: this work will – in sha'Allah – provide the Muslim with a deep grounding in the three essential sciences of the Deen, enabling him or her to identify deviant doctrines that promote an anthropomorphic conception of Allah, or discredit the schools of fiqh, or deprive the Deen of its spiritual teachings.

Shaykh Ali Laraki

6 See in this regard *Maraqi as-Su'ud* by ash-Shinqiti, verse 958 and its various commentaries *Nashr al-Bunud*, *Nathr al-Wurud*, and others.

TRANSLITERATION AND ITALICISATION OF WORDS

In this work we have adopted the following procedure with respect to the transliteration and italicisation of words. The words are not transliterated according to the standards of academic orientalists as in our view this is a part of a quite concerted programme of erasing the Arabic script and replacing it with the Latin script, as has been done successfully in Turkey, Malaysia and many other countries. The serious student will easily find out the Arabic spellings of the words we have used. For the person whose concern is simply to know the knowledge outlined in this book for the sake of his own practice, the transliteration scheme we have adopted will be quite sufficient.

With italicisation we have followed the convention that foreign words and book titles are to be italicised, but have taken the position that a huge number of Arabic words are no longer foreign words in the English language, particularly for Muslims, but indeed even for non-Muslims. For the purposes of this book words such as Sunna, wudu', salat and zakat are to all intents and purposes now English words.

VERSE NUMBERINGS

Throughout the book, the verses to which the commentary refers are indicated numerically in the margins. The poem itself can be found on p.155.

Sidi 'Abd al-Wahid ibn 'Ashir's Introduction

THE AUTHOR, ABDULWAHID ibn Ashir,[7] starts with his own name, 1
because the name of an author is an important matter, as it is not permissible to act upon or give fatwa from a book whose author is unknown and whose content is not confirmed.

He mentions the name of Allah in accordance with the hadith: "Any matter of importance not started with the name of Allah will be severed." (Abu Dawud, an-Nasa'i and Ibn Majah from Abu Hurayra).

He starts by praising Allah (*hamd*) emulating the Qur'an which begins with 2
praise of Allah, and in accordance with the hadith that says: "Any matter of importance not started with the praise of Allah will be mutilated." (Abu Dawud from Abu Hurayra.)

The knowledge we are responsible for is that without which we could not carry out wudu', ghusl, salat, fasting, paying zakat if we have the means, hajj if we are able, and everything related to what is obligatory to believe about Allah and His Messenger ﷺ.

The same relates to sale, trade partnerships (*qirad*) or rent for whoever undertakes them. It is obligatory by consensus to learn the rulings of any transaction that one wants to undertake, since it is forbidden to undertake any action until Allah's rulings are known related to that undertaking.

The author asks Allah for blessings and peace (*as-salatu wa as-salam*) upon 3
Muhammad, his family, companions and the one who follows him. Asking for blessings upon the Prophet ﷺ is obligatory once in a lifetime if one is able. According to another opinion it is considered a sunna *mu'akkada* (confirmed sunna) which should not be left, except by the one who has no good in him.

---

7 See a short biography of him in the Preface.

According to another opinion, it is obligatory every time he is mentioned ﷺ.

There are different opinions as to whether asking for blessing (*as-salat*) is allowed for other than the prophets, but it is definitely allowed in the statement "May Allah bless Muhammad, his family and companions." His family are his believing relatives from Banu Hashim and descendents. A Companion (*Sahabi*) is one who met the Prophet ﷺ and believed in him.

The Prophet ﷺ is the great means (*wasita*) through which every good has reached us and will ever reach us.

4 The author asks for help from Allah, the Glorious, in composing verses which will benefit the unlettered. The definition of the Glorious (*al-Majid*) is the One Who has reached the uttermost honour, perfection and expansion in the Kingdom, beyond which no more is attainable.

The definition of unlettered (*ummi*) is someone who is intellectually in the condition he was at birth, unaffected by reading or writing.

5 The author explains that his poem is about:

1. The tenets of belief ('aqida) of al-Ash'ari
2. The jurisprudence (fiqh) of Malik and
3. The path (tariqa) of al-Junayd, the Wayfarer

Abu'l-Hasan al-Ash'ari died around 332 AH.[8] He is the imam of the science of 'aqida, which deals with the tenets of belief. He was the first one who wrote extensively about it, summarising and defending it from deviation. The other two correct schools are the Maturidi and the Athari.[9]

Imam Malik is the 'Imam of Madina, the Abode of the Hijra' from the third generation of the salaf, who passed away in the year 179 AH.[10] His school is unique in representing the living Prophetic sunna as mass-transmitted by

8 Approximately 943 CE.

9 Beware of some Hanbalis who claim to follow the Athari school but who, as Ibn Juzayy says in his *al-Qawanin al-Fiqhiyyah*, have fallen into anthropomorphism by literal interpretation of some *ayats* of the Qur'an and some hadith. An example of an authentic *athari* formulation of the 'aqida is the chapter on beliefs of the *Risala* of Ibn Abi Zayd al-Qayrawani.

10 Approximately 795 CE.

the inhabitants of Madina al-Munawwara, the blessed city where the Shari'a was revealed. He is known as the Imam of the Imams, as he taught Imam Muhammad ash-Shaybani al-Hanafi, Imam ash-Shafi'i and, via ash-Shafi'i, Imam Ahmad. He also taught the leading hadith transmitters of his day who in turn transmitted his hadith to al-Bukhari, Muslim and the other Imams of hadith narration.

Imam al-Junayd died in the year 297 AH[11] in Baghdad and was known as 'the Imam of the World of his Time.'[12] Though an outstanding scholar of fiqh, he is best known as the Imam of the Sufis. He formalised the prophetic science of self-purification and Ihsan for subsequent generations and many authentic tariqas trace their lineage and method to him.

The author has hereby informed us that his poem has gathered the most important elements of the three sciences, 'aqida, fiqh and tasawwuf, which relate to the three parts of the Deen: Iman, Islam and Ihsan.[13]

11 Approximately 909 CE.

12 According to the Muslim historian Ibn al-Athir.

13 From *al-Habl al-Matin* by al-Marrakushi.

Beliefs

An Introduction to the Book of Tenets of Belief which Help its Reciter Obtain the Objective

INTRODUCTION TO THE ʿAQIDA (BELIEFS)

ʿAQIDA IS THE science of Iman. The word comes from the root *ʿayn-qaf-dal*, ʿto bind together.'

For more than a thousand years the Ahl as-Sunna wa'l-Jamaʿa have been defined as those following the Ashʿari, Maturidi or Athari schools of ʿaqida.

This book deals with the largest school, that of Imam al-Ashʿari, born in Basra in 260 AH.

His ʿaqida was formulated to counter deviant sects, primarily the Muʿtazili, who put reason above the Qur'an and the Sunna. The Imam was himself a Muʿtazili until seeing the Messenger ﷺ in a dream, after which he turned on them using their own logic.

Allah willing, this section gives the rational evidence needed to prove the existence of Allah and to ward off false doctrines, such as Allah being in His creation, whether incarnate in a human body, in the sky or in nature – may He be exalted beyond anything they describe!

The science of ʿaqida is traditionally divided into three areas:

1. Knowledge of Allah and His Attributes, or theology (*ilahiyyat*)
2. Knowledge of the Messengers and their attributes, or prophetology (*nubuwiyyat*)

3. Knowledge of the Unseen, i.e. extrasensory matters (*ghaibiyyat*). It includes eschatology[14] and angelology,[15] etc.

And sometimes a fourth is added:

4. Knowledge of miscellaneous matters (*jami'*), such as who the Rightly Guided Caliphs (*al-Khulafa ar-Rashidun*) were, the preconditions of the Caliphate, and other subjects

'Aqida is also called Kalam, Tawhid or Usul ad-Deen. *Kalam* is the name of the method, *Tawhid* deals with the unity of Allah and *Usul ad-Deen* refers to the 'foundations of the religion,' as no action has a foundation or value without a correct creed.

THE RATIONAL PROPOSITION AND ITS PARTS

6 Propositions are divided into three kinds. The science of 'aqida revolves around the first and the third kind:

1. Rational propositions (*hukm 'aqli*), which are confirmed or rejected by the intellect, such as: 2+2=4, and do not need verification using the following two categories:
2. Empirical propositions (*hukm 'adi*), which are confirmed or rejected by experience, such as: it hurts to put your hand into an open flame
3. Religious propositions (*hukm shar'i*), which are confirmed or rejected by the Qur'an or Sunna, such as: there is life after death

7-9 Rational propositions divide into three kinds:

1. The necessary proposition – which must necessarily be true, such as: a body is either moving or still
2. The inconceivable proposition – which cannot possibly be true, such as: a body may move and be still at the same time
3. The conceivable proposition – which may or may not be true, such as: a body may be moving

Each of these three types further divides into two:

14 The study of subjects such as death, the Last Day, the Judgement, the Garden and the Fire.
15 The study of angels.

- □ Propositions that may be instantly understood (*daruri*), such as:
  - The necessary rational proposition: 1+1=2
  - The inconceivable rational proposition: 1+1=3
  - The conceivable rational proposition: a body may be moving
- □ Propositions that need reflection (*nadhari*), such as:
  - The necessary rational proposition: ½ of a ¼ of a $^1/_{10}$ of 80 is 1
  - The inconceivable rational proposition: ½ of a ¼ of a $^1/_{10}$ of 240 is 2
  - The conceivable rational proposition: Allah will punish the obedient slave

RESPONSIBILITY AND ITS SIGNS

The first obligation on the responsible human (*mukallaf*) who has the **10-11**
capacity to search and reflect is to know:

1. Allah and
2. His Messengers by the attributes of both, for which He has established proofs and evidences

Knowledge (*'ilm*) is the opposite of *taqlid*[16] in this context of 'aqida. *'Ilm* is defined as 'firm conviction, based on a proof, which corresponds to reality'.[17]

Therefore, whoever has faith (*Iman*) based on something other than firm conviction (*jazm*), such as speculation (*dhann*), doubt (*shakk*) or illusion (*wahm*), his *Iman* is invalid by consensus. It is the same if faith does not correspond with reality, like the faith of Christians, Jews and atheists.

As for having faith not based on proof (*taqlid*), the widespread opinion of the 'ulama of *Ahl as-Sunna wa'l-Jama'a* is that it is not valid to rely on it in matters of belief. It is therefore necessary that the believer base his belief on solid proof, not just following an opinion.

16 *Taqlid* is defined as following somebody's opinion without knowledge of the proof on which it is based.

17 Imam Malik has also defined *'ilm* as a light Allah puts in the heart.

Unlike fiqh and tasawwuf – where *taqlid* is necessary[18] – *taqlid* is not acceptable in matters of 'aqida.

12-13 There are three preconditions of being *mukallaf*:

1. Sanity
2. Puberty, recognised by either
   - □ First menstruation
   - □ Pregnancy
   - □ Emission of reproductive fluid (as during wet dreams)
   - □ Thick pubic hair or – if none of these appear –
   - □ Being 18 lunar years old – about which there are two valid opinions:
     - Entering the 18th lunar year (after completing the 17th) or
     - Completing the 18th lunar year
3. Access to the message of Islam[19]

18 In tasawwuf, the word *musahabah* (following or accompanying) is used rather than *taqlid*. Nevertheless it conveys the same meaning with the difference that in fiqh it is allowed to follow a dead imam whilst in tasawwuf the imam or shaykh has to be alive (see the *Diwan Bugyat al-Muridin as-Sa'irin* by Shaykh Muhammad ibn al-Habib al-Amghari al-Idrisi p. 70, Dar as-Sadir, Beirut).

19 As have the majority of people today. If such people out of arrogance, negligence, stubbornness or fanaticism do not consider Islam, they are kuffar, and Allah knows best.

The Source of the Principles and Their Contained Beliefs

THE PRIMARY PRINCIPLE that the author mentions here is what constitutes the source and basis of the other four principles mentioned in the work and its condition of validity:[20] the *Shahadatayn* or Double Testimony, which is to bear witness that there is no god but Allah and Muhammad is the Messenger of Allah ﷺ.

WHAT IS NECESSARILY TRUE REGARDING ALLAH

The following thirteen statements about Allah and His Attributes are necessarily true: **14-16**

1. Allah has **Existence** (*Wujud*) – the opposite is rationally inconceivable, as discussed in the section on the proofs of Allah's Attributes. Allah is the Being Who necessarily exists.
2. Allah has **Pre-Eternity** (*Qidam*) – Allah has always existed without a beginning.
3. Allah has **Everlastingness** (*Baqa'*) – His existence is continuous and permanent, i.e. Allah will never come to an end.
4. Allah has **Absolute Independence** (*Ghina Mutlaq*), also termed **Self-subsistence** (*Qiyam bi'n-Nafs*), which has two aspects:
   - □ He does not need any other thing to provide Him with existence.
   - □ He is not an attribute needing an essence in which to subsist.[21]

---

20 The author makes another reference to this in verse 42.

21 Christian doctrine entails belief that Divinity is an attribute that can subsist in a human essence and that therefore a human can be divine, which is inconceivable because Divinity is not an attribute but an essence.

5. Allah has **Difference from Creation** (*Mukhalafatu li'l-hawadith*) – i.e. He is different without a similitude; nothing resembles Allah, whether in essence, attributes or actions.[22]
6. Allah has **Oneness** (*Wahdaniyya*) in His Essence, Attributes and Actions,[23] which has two aspects:
   - □ His Essence is not composed of parts.
   - □ There is no other being that has an essence, attributes or actions like His Essence, Attributes or Actions.
7. Allah has **Power** (*Qudra*) – the capacity to bring into existence or bring to an end anything contingent (conceivable), meaning anything that can conceivably exist.
8. Allah has **Will** (*Irada*) – the capacity to particularise every contingency with what He wants, in that, He decides the moment of appearance, colour, weight, size, movement, etc. of any given thing or event.
9. Allah has **Knowledge** (*'Ilm*) – Allah knows everything that is necessary, conceivable and inconceivable. He knows what was, what is and what will or will not be.
10. Allah has **Life** (*Haya*) – this is a condition for the existence of Power, Will and Knowledge.
11. Allah has **Hearing** (*Sam'*) – the ability to hear all things without exception or restriction, unlike human hearing, which is dependent on an apparatus (the ear), air pressure, frequency, volume and range.
12. Allah has **Sight** (*Basar*) – the ability to see all things without any

---

22 Beware of those who interpret some verses of the Qur'an or some passages from the hadith literally as in the case where Allah speaks about His hand or His face, or about coming or descending. Those who say that the hand, the face, the descending and the coming of Allah are literal, are comparing Allah – *subhanahu wa ta'ala* – to His creation and are committing anthropomorphism, even if they try to explain it by saying that His hand is not like our hands and adding the expression, "as befits His Majesty". See the commentary of *Jawharat at-Tawhid* by al-Baijury, verse 40. The correct position is to believe in the Qur'anic verse or hadith without taking it literally and without interpreting it, or, alternatively, to interpret it in a correct way as some of the later 'ulama of *kalam* have done.

23 This is the correct way to understand the Oneness of Allah and not the modern version that divides the Oneness of Allah into Oneness of Lordship, Oneness of Divinity and Oneness of the Names and Attributes (*Tawhid ar-Rububiyyah*, *Tawhid al-Uluhiyyah* and *Tawhid al-Asma' wa's- Sifat*). For the reasons against this kind of division of the Oneness of Allah, see *Assessment of the Division of Tawhid into Uluhiyyah and Rububiyyah* (*Naqd Taqsim at-Tawhid ila Uluhiyyah wa Rububiyyah*) by Abu Mahasin Jamal ad-Din Yusuf bin Ahmad ad-Dijwi al-Maliki al-Azhari.

exception or restriction, unlike human sight, which is dependent on an apparatus (the eye), light, colour, size, range, etc.

13. Allah has **Speech** (*Kalam*) which means:
    - □ His eternal Attribute of Speech (Self-Speech, *al-Kalam an-Nafsi*), present in the Essence of Allah, which is unlike human speech – not being in any language, nor made up of letters, sounds or silences and not constricted in time
    - □ His revealed Books, such as the Injil and the Qur'an, also known as His Verbal Speech (*al-Kalam al-Lafdhi*), are called His Speech for two reasons:
        - They indicate, in various human languages such as Hebrew or Arabic, some of the semantic import of the infinite meanings encompassed in the the eternal Attribute of Speech.
        - They are the Word of Allah, not of anyone else.

The word *Qur'an* indicates both the eternal Attribute of Speech and the revealed Book. Therefore, to avoid confusion between *al-Kalam al-Lafdhi* and *al-Kalam an-Nafsi,* it is forbidden to say that the Speech of Allah or the Qur'an is created.

The Attributes divide into four distinct categories:

1. The first Attribute – Existence – is called 'the Essential Attribute' (*Sifa Nafsiyya*) because it refers to the Essence itself without any addition.
2. The following five – Pre-Eternity, Everlastingness, Absolute Independence, Difference and Oneness – are called 'the Negative Attributes' (*Sifat Salbiyya*) because they negate what is not proper for the Divine Essence.
3. The last seven – Power, Will, Knowledge, Life, Hearing, Sight and Speech – are called 'the Attributes of Meaning' (*Sifat al-Ma'ani*) because these Attributes, contrary to those from the previous two categories, convey additional meanings about the Essence.
4. Added to these seven are another seven Attributes derived from these meanings (*Sifat Ma'nawiyya*) which mean that Allah is:
    - □ Powerful
    - □ Willing
    - □ Knowing
    - □ Living

- ☐ Hearing
- ☐ Seeing
- ☐ Speaking

WHAT IS INCONCEIVABLE REGARDING ALLAH

17-19 It is inconceivable that Allah should be subject to the opposites of the aforementioned thirteen[24] Attributes, which are:

1. Non-existence (*'adam*)
2. Beginning[25] (*huduth*)
3. Extinction (*fana'*)
4. Dependence (*iftiqar*)
5. Resemblance (*mumathala*)
6. Plurality– i.e. to have a partner or consist of parts (*ta'addud*)
7. Incapacity (*'ajz*)
8. Compulsion (*karaha*)
9. Ignorance (*jahl*) – and this includes doubt and forgetfulness
10. Death (*mawt*)
11. Deafness (*samam*)
12. Blindness (*'ama*)
13. Muteness (*bakam*)[26]

WHAT IS CONCEIVABLE REGARDING ALLAH

20 Allah may do anything that is conceivable or leave it undone in non-existence. 'Conceivable' here means anything 'rationally possible',[27] because He will not create a god like Himself or cause His Being to cease, for that is rationally inconceivable.

24 Or twenty Attributes, if we take into consideration the *sifat al-ma'nawiyyah.*

25 Or contingency.

26 We should add the opposite of the *Sifat al-Ma'nawiyyah* which are to be: incapable, compelled, ignorant, dead, deaf, blind and mute.

27 We may also say contingent.

THE PROOF OF HIS EXISTENCE

His existence has a conclusive proof: the necessity of a Creator for everything that is created, here meaning anything which is contingent,[28] because, if the universe began by itself, then two contradictory statements would both be true: **21-23**

1. Balance between the possibility of existence or non-existence of anything in particular or the universe in general.[29] The existence of the universe in general or of a single event in it in particular is not necessary.
2. Preponderance (overbalance) of existence, which is an empirical fact (i.e. the universe entered existence and its non-existence ceased, its existence overbalanced its non-existence).

But it is inconceivable that 1 and 2 should occur together. It is contradictory for the existence of the universe to be balanced and in equilibrium with its non-existence and to be, at the same time, overbalanced so as to make the universe emerge into existence. In other words, the universe does not possess the intrinsic capacity of making itself appear into existence. Therefore, an outside agent must be necessary in order to push the balance in favour of existence, for it is inconceivable that one of two equals, by themselves, could overbalance the other.

The beginning (*huduth*) of the universe is derived from the continuous beginning and ending of accidents paired with the knowledge that accidents are inherent to essences, (e.g. the attributes of 'rest' and 'motion' are inseparable from objects).[30]

PROOFS OF THE REST OF HIS ATTRIBUTES

So if Pre-Eternity were not His necessary Attribute, then His having a beginning would obligate one of two impossibilities: **24**

---

28 See verse 9 and its commentary for a definition of 'conceivable'.

29 And this is what we mean when we say that something is 'contingent'.

30 These changes prove that the universe is in the flow of time. Any conceivable thing must exist in the flow of time, i.e. have a beginning. If its existence had no beginning, it would be necessary, instead of merely conceivable.

1. A vicious cycle which is a circle of creators such that A creates B, B creates C, and C creates A, which is inconceivable since C's existence is conditional on A's prior existence; or
2. An eternal regress back in time ending in the present, which is inconceivable because an infinity that ends is not infinite, since an actual infinite amount of time can never be traversed[31]

25 If He had no Everlastingness, then His Attribute of Pre-Eternity would be nullified, because what has no beginning can never end.[32]

If He were similar in any way to originated things, then His having a beginning would definitely be true, as He would also be originated.

26 If Independence was not His necessary Attribute, then He would be in need of either:

1. A maker (*mukhassis*), the impossibility of which has been dealt with above; or
2. A locus (*mahall*), i.e. an essence in which to subsist. Allah would then be an attribute (such as hardness is the attribute of stone) which is inconceivable as He is necessarily described by the Attributes listed here, and an Attribute is not described by an attribute.

If He did not have Oneness, then He would not be able to create the world, because if there were more than one God, one of three cases would occur:

1. Two equal gods would will different events and one would have his way, in which case the other would be powerless i.e. not a god; but since they are equal, this would be true for both of them.
2. Two different gods would will different events and one have his way, then one would be powerless and the other would be the One God.
3. Two gods would will the same event, which requires that two wills make an impression on an indivisible thing – i.e. the simplest particle.

---

31 An actual infinite is a set or series that has an infinite number of members, such as all positive real numbers {1,2,3...}: that set actually contains an infinite number of members. So the statement "The universe had no beginning" is rationally inconceivable because it claims an infinite amount of time having passed during the history of the universe to find a limit in the present moment.

32 If He had no Everlastingness, He would be subject to extinction, His Existence being therefore conceivable, not necessary. But we have been able to prove that the existence of Allah is not merely conceivable, but necessary.

Only one will can make an impression at any given time.[33] Otherwise it would be like two blacksmiths whose hammers hit the same point on the same piece of steel at the same time, which is inconceivable.

If He did not possess Life, Will, Knowledge and Power, then you would not see a created world. This is because the 'ulama' of *Ahl as-Sunna wa'l-Jama'ah* have resolved that the effect of the eternal power depends on Him willing it to take place. Nothing happens or does not happen except by His Power and Will. The willing of anything depends on His Knowledge of it, because nothing can be intended unless it is known. Furthermore, Power, Will and Knowledge depend on Life, since Life is the precondition of these three Attributes.[34] Therefore, the existence of any effect or being depends on its maker being described by the four aforementioned Attributes. In other words, the mere existence of the universe is a proof that its Maker possesses Life, Will, Knowledge and Power. **27**

The consequences of the six arguments are definitely false; therefore the antecedents are likewise false. **28**

The results of the above six logical arguments (the underlined if/then clauses) are false, so it follows that their propositions are also false.[35]

Hearing, Sight and Speech have two types of proof: **29**

1. From the revealed statements (*naql*):
   - ☐ Allah's Words: *He is the Hearing, the Seeing.*[36]
   - ☐ *Allah spoke to Musa.*[37]
   - ☐ The hadith: "Be gentle with yourselves, you are not calling someone who is deaf and absent, you are calling One Who is

---

33 Hence the atomism of the Ash'aris, the concept of which is supported by modern physics, although the term 'atom' used by the Ash'aris is different from what the physicists mean by 'atom'.

34 This sequence of dependencies is not a process that takes place in time. There is no sequence in Allah's qualities, except in human comprehension.

35 This syllogism is called in classical logic, *modus tollens* (Latin for 'the way that denies by denying'). It has the following argument form: If P then Q. No Q therefore no P. It can also be referred to as denying the consequent.

36 Al-Baqara, 137.

37 An-Nisa', 164.

Hearing, Seeing."[38]

2. Rationally they also properly befit His perfection. It is not conceivable that the creature is hearing, seeing and speaking, while its Creator is deaf, blind and dumb.

NB: There is consensus (*ijma'*) that Allah is described by these thirteen Attributes, though some may be proved by intellect alone, and others need the support of textual evidence, such as Hearing, Sight and Speech, as discussed.

30 The proof that Allah may or may not do anything that is conceivable, is that if doing anything contingent were inconceivable or necessary for Allah, then that thing would be necessary and conceivable at the same time, thereby contradicting and therefore inverting the qualification of realities.

WHAT IS NECESSARILY TRUE REGARDING THE MESSENGERS

31 The noble Messengers are necessarily described by the following attributes:

1. Truthfulness (*Sidq*), which relates to their words: everything that they say is true and corresponds to reality.
2. Trustworthiness (*Amana*), which relates to their actions: they are protected by Allah from doing anything forbidden or disliked, outwardly and inwardly.
3. Conveyance of the message (*Tabligh*), which means that they deliver the entire message, not adding or leaving anything out, willingly or unwillingly.[39]

WHAT IS INCONCEIVABLE REGARDING THE MESSENGERS

32 The opposites are inconceivable for them:

38 Transmitted by Abu Musa al-Ash'ari and recorded by al-Bukhari and Muslim.

39 Some 'ulama of *kalam* add a fourth attribute, the attribute of 'Sagacity' (*Fatana*) which means the necessary alertness to refute the arguments of their opponents. See *Jawharah at-Tawhid* (verse 59) by Imam Ibrahim al-Laqqani.

1. Lying by saying something that does not conform to reality
2. Deception by doing something forbidden or disliked
3. Concealing the message by neglecting to say anything that they were ordered to transmit to creation[40]

WHAT IS CONCEIVABLE REGARDING THE MESSENGERS

Normal human conditions are conceivable for them, such as: 33

- Illness (not leading to deficiency, such as leprosy, madness, etc.)
- Hunger and thirst
- Pain
- Being physically harmed by people
- Eating and drinking
- Marrying
- Forgetting things they have not been ordered to convey

None of the conditions that may affect them can constitute a belittlement of their lofty station, such as making them appear vile or repulsive to people.

PROOFS OF THE ATTRIBUTES OF THE MESSENGERS

The proof that they are truthful is that if they were not, it would necessarily 34
imply that Allah had lied in confirming them by His miracles (*mu'jizat*). His miracles are recognised by four traits:

- Being abnormal events
- Coupled with the claim of messengership
- Used to challenge the opponents of the message before they occur
- Being impossible to replicate

The miracle is as if Allah says: "This servant is telling the truth in all he 35
conveys", which is a confirmation of that Messenger. If his words were not true, Allah would be confirming a lie and confirming a lie is also to lie, and it is inconceivable for Allah to lie.

The proofs of trust and delivery of the message are identical as mentioned 36

40 Likewise, the opposite of sagacity is dullness, i.e. not being able to establish the proofs and contest the arguments of their opponents.

below:

The proof that they delivered the message and were trustworthy is that Allah has ordered us to follow them, so if they were treacherous, disobedience would become obedience.[41]

37 The proof that they experience what people are exposed to is the mass (*mutawatir*) transmission of eyewitnesses to their occurrence. Part of its wisdom is to console them and humanity, as well as informing them about the low value of this world (*dunya*), as Allah withholds it from His Messengers.

THE SHAHADA

38-39 The Shahada, the statement, "There is no god but Allah, Muhammad is the Messenger of Allah" in Arabic is:

La ilaha illa Allah, Muhammadun rasulullah.

The Shahada contains all the meanings described above and it includes everything conceivable, inconceivable and necessarily true concerning Allah and His Messengers. For this reason it is the motto of Iman. The term *Ilah* (God) means, "He Who is independent of everything and upon Whom everything depends." Therefore, if we replace the word *Ilah* from the Shahada with its definition, it will be: "There is no one independent of everything except Allah, on Whom everything else depends, Muhammad is the Messenger of Allah."

Allah being described as 'He Who is independent of everything' (independence – *istighna'*) includes the following twenty-eight tenets of belief:

1. His Existence (*Wujud*)
2. His Eternity (*Qidam*)
3. His Everlastingness (*Baqa'*)
4. Absolute Independence (*Ghina al-Mutlaq*)

41 As for the proof that they have acute intelligence it is that, without it, they would be unable to establish an argument against their opponents. However, the Qur'an indicates that they must establish arguments, which is only possible with acute intelligence.

5. Difference from His creation (*Mukhalafatu li'l-Hawadith*)[42]
6. Hearing (*Sam'*)
7. Sight (*Basar*)
8. Speech (*Kalam*)[43]
9. His being Hearing (*Sami'*)
10. His being Seeing (*Basir*)
11. His being Speaking (*Mutakallim*)
12. His being free from objectives (*Munazzah 'ani'l-aghrad*)[44]
13. Not being obliged to do something or to leave it undone (*'adam wujub fi'l shay' 'alayh aw tarkih*)[45]
14. Nothing else having an effect through intrinsic power[46] (*kawn ash-shay' mu'aththiran bi'l-quwwa*)

To these fourteen beliefs, we should add the impossibility of their opposites.

As for Allah being 'He on Whom everything depends' (dependence – *iftiqar*) it includes the following twenty-two tenets of belief:

1. His Power (*Qudra*)
2. His Will (*Irada*)
3. His Knowledge (*'Ilm*)
4. His Life (*Haya*)[47]

42 If Allah did not have these five Attributes, He would have been in need of a creator, which would imply a vicious circle or an eternal regression.

43 Without the Attributes of Hearing, Sight and Speech, He would be deficient and, therefore, in need of someone to remove these defects from Him.

44 This means that Allah has no objective or need to be attained through act or law. If that was the case, then He would need to fulfil that objective which could only be completed through His creation. This is a deficiency incompatible with Allah's Independence.

45 Otherwise, it would be obligatory on Him to do something and He would be in need of that thing in order to attain perfection. So, for example, reward is a grace from Him and not anybody's right. "*He is not questioned as to what He does, but others are questioned.*" (Surat al-Anbiya' 21:23)

46 This means that things would have an effect by a power created in them by Allah. If that was so, then Allah would be in need of intermediaries in order to act (a fire would not burn and a knife would not cut except by an intrinsic power given to them by Allah), and to affirm such a belief is a deviation by consensus. Rather, Allah creates both the fire and its burning and the knife and its cutting, the burning and the cutting by convention being attributed to the fire or the knife.

47 If Allah did not have these four Attributes, then He would not have been able to create anything and nothing would depend on Him.

5. His being Powerful (*Qadir*)
6. His being Willing (*Murid*)
7. His being Knowing (*'Alim*)
8. His being Living (*Hayy*)
9. His Oneness (*Wahdaniyya*)[48]
10. Nothing else having an intrinsic effect by its nature[49] (*kawn ash-shay' mu'aththiran bi't-tab'*)
11. The created nature of the universe (*huduthu'l-'alam*)[50]

To these eleven beliefs, we should add the impossibility of their opposites. Therefore, 'there is no god but Allah' includes fifty beliefs in total.

Our saying "Muhammad is the Messenger of Allah" ﷺ includes sixteen beliefs:

1. In the rest of the Prophets and Messengers (*Anbiya' wa Rusul*)[51]
2. In the Angels (*Mala'ika*)
3. In the Books (*Kutub*)
4. In the Last Day (*al-Yawm al-Akhir*)[52]
5. The Prophets' Truthfulness (*Sidq*)
6. Their Trustworthiness (*Amana*)
7. Their Conveyance (*Tabligh*)
8. The possibility of their being subjected to human conditions (*jawaz al-a'rad*)[53]

To these eight we should add the impossibility of their opposites, bringing the total contained in the Shahada to sixty-six beliefs, allowing us to realise that despite the few words it contains, the Shahada includes what is necessary for the responsible person to know and believe regarding Allah and His Messengers.

---

48 If Allah was not one, the multiple gods would be incapable of creating the world and nothing would be dependent on Him.

49 This means affirming that things have an effect on other things by themselves. If things could have effects on others by themselves, then effects would be independent from Allah and would not need Him to occur. Affirming such a concept is disbelief by consensus.

50 If the world or part of it was not created and did not have a beginning, it would be independent of Allah.

51 The number of prophets is 124,000 and the number of messengers is 314. The first of them is Adam and the last of them Muhammad – may Allah bless and grant peace to all of them.

52 This is because whoever believes in Muhammad ﷺ has to believe in everything he has informed us of regarding the unseen and the Next Life.

53 The proofs of 5 to 8 have been duly explained in verses 34 to 37.

The Shahada is the greatest dhikr so reflect on it, occupy your life with it[54] and you will accumulate great treasure for this life and the next.[55] Neither Iman nor Islam is accepted with any other statement. 40

The Messenger ﷺ said: "The best thing the Messengers before me and I myself have said is,'*la ilaha illa Allah wahdahu la sharikalah* (There is no god but Allah alone without partner).'"[56]

ISLAM AND ITS PILLARS

Complete Islam is to fulfil the commands and avoid the prohibitions of Allah with the seven limbs: the eyes, ears, tongue, hands, feet, genitals and stomach. Partial compliance is either: 41

1. Deficient (*naqis*) Islam, such as performing what is obligatory while not avoiding what is forbidden or
2. *Kufr* – such as a person denying the five salats as a pillar of Islam

However, the judgement of a person being Muslim is confirmed externally only by his uttering the Shahada.

The pillars of Islam are five, the first of which conditions the validity of the other four: 42-43

1. Shahada: Uttering, understanding and believing both parts of the dual statement:

   Ashhadu an la ilaha ill'Allah wa ashhadu anna Muhammadan rasulullah – I bear witness that there is no god but Allah and I bear witness that Muhammad is the Messenger of Allah

2. Salat: Performing the five daily salats as they should be done
3. Zakat: Due on gold, silver and certain crops and livestock
4. Fasting (*Sawm* or *Siyam*): Fasting the month of Ramadan
5. Pilgrimage (*Hajj*): Journeying to the Masjid al-Haram in Makka and the Plain of 'Arafa if able

The rules of Islam also encompass all other aspects of life.

54 Until its meaning becomes mixed with your flesh and bones.
55 And you will witness endless secrets and wonders.
56 Recorded by Malik in his *Muwatta*.

IMAN

44-45 Iman is firm conviction in:

1. Allah, the One God possessing all the above mentioned Attributes of Majesty and Perfection as they befit Him. He is the only Creator, with no partner in either:
   - □ His Divinity or
   - □ His deserving worship

   Nothing happens in His Kingdom except what He wills, but actions of obedience happen by His will and orders, whereas actions of disobedience happen by His will and against His orders.

2. His Angels (*Mala'ika*) who are His slaves, incapable of disobedience, made of light, who do not eat or sleep and are not male or female. Some act as messengers of Allah to humans, others have other functions such as recording men's actions. They are very numerous.
3. His Books (*Kutub*) which are His Endless Speech in His Essence with no beginning, no words and no letters, revealed to some Messengers by created words on tablets or on the tongues of angels. Everything in them is true.

   Generally we believe in a total of one hundred and four books, and specifically we believe in four:

   - □ The Tawrah (Torah), revealed to Sayyidina Musa (Moses) عليه السلام
   - □ The Zabur (Psalms), revealed to Sayyidina Dawud (David) عليه السلام
   - □ The Injil (Gospel),[57] revealed to Sayyidina 'Isa (Jesus) عليه السلام
   - □ The Qur'an, revealed to Sayyidina Muhammad ﷺ, which abrogates and preserves all previous revelations
4. His Messengers (*rusul*). Altogether we believe in 314 messengers (who brought new Shari'as), and 124,000 prophets (confirming previously revealed Shari'as). Specifically we believe in all those mentioned in the Qur'an.

   His Messengers were sent to His servants to guide them and to complete and perfect their lives in this life (*dunya*) and in the next

57 Today's versions of the Torah, Psalms and Gospel have been tampered with by the followers of Judaism and Christianity and do not coincide with the original revealed ones.

(*akhira*). They were supported by miracles and showed the responsible person (*mukallaf*) what to do. We respect all and make no difference between them. Allah has protected them against imperfection and wrong actions, great or small, before and after their messengership.

It is obligatory to believe in any Messenger, Book or Angel named in the Qur'an or the Sunna, and denying any of these is disbelief (*kufr*).

5. The Rising (*Ba'th*) from the dead on the Last Day (*Al-Yawm al-Akhir*) in body and soul, and that it is imminent.
6. The Decree (*Qadar*) – the sweet and bitter of it – which Allah has written before time. What is written will happen, what is not written will not.[58]
7. The Bridge (*as-Sirat*) across the abyss of Hell to be crossed by people in a manner identical to their conduct in the dunya; some will cross to safety like lightning, some like the wind, some like racehorses, while others will fall.
8. The Scales (*Mizan*) with an indicator and two pans, one for the good actions and one for the bad actions, which will be filled with either the sheets listing people's actions, or bodies that Allah creates to represent them.
9. The Pool (*al-Hawd*) of the Messenger ﷺ which is whiter than milk and sweeter than honey. The goblets around it are more numerous than the stars, and whoever drinks from it shall never be thirsty.
10. The Garden and the Fire (*Al-Jannah* and *an-Nar*), which are creations of Allah, existing and endless. Three points should be noted:
    - Allah will grant those who enter the Garden the vision of His Noble Face.
    - Adam عليه السلام, His Prophet and Khalifa, descended from the Garden to the earth.
    - He created the Fire as the eternal abode for those who disbelieve in Him, His books, His Messengers and will prevent them from seeing His Face.

58 The above are known as 'The *Six Pillars* of Iman'.

IHSAN

46-47 The word Ihsan has two lexical meanings in the Arabic language:

1. Acting with excellence, as it should be done, which is intended here, and
2. To do good to people

Ihsan in this first sense is the excellence of Islam and Iman and it consists of two levels, the first being the highest and the second basic:

1. That you contemplate the Real (*Al-Haqq*) with your heart, as if you see Him with your eyes (*mushahada*) and
2. If you do not, then realise that the Real sees you.

The science of Ihsan is discussed in detail in the Book of Tasawwuf.

THE DEEN

We should know that the Deen consists of three parts mentioned in the Hadith of Jibril ﷺ:

1. Islam
2. Iman
3. Ihsan

Therefore, a person's Deen is faulty if it lacks any of these three elements. So take hold of these three with an iron grip.

The proof of this is in the end of the Hadith with the words of the Messenger of Allah ﷺ "That was Jibril who came to teach you your Deen," and Imam al-Bukhari comments that, "He made all of that the Deen."

Introduction to the Science of the Sources of Fiqh

An Introduction to the Principles of Jurisprudence which Help Reach Understanding of its Branches

INTRODUCTION TO FIQH

FIQH IS DEFINED as 'the science of the knowledge of the judgements derived from the Qur'an and Sunna regarding particular actions of responsible people' (*mukallaf*). Fiqh covers two areas:

Acts of worship (*'ibadat*), that deal with the relationship between the responsible person and Allah, such as salat, fasting and hajj.

Transactions (*mu'amalat*), that deal with the relationship between people, such as trade, marriage, penal law, inheritance, etc.

TYPES OF JUDGEMENT

Linguistically *hukm* is a judgement about something or somebody. A **48-49**
judgement in the Shari'a (*hukm shar'i*) is the Speech of Allah which determines the conduct of the responsible person and can only be confirmed by revelation, not rationally or empirically.

The types of judgement are as follows:

1. A request (*taklif*), which may be either
   - □ Permission (*idhn*) to act or a
   - □ Demand (*talab*) – which may be strict (*jazim*) or not (*ghair jazim*) – to either:
     - Perform an action (*fi'l*) or
     - Refrain from an action (*tark*).
2. A stipulation (*wad'*), which is of three kinds:

- ☐ Cause (*sabab*) – as illustrated in the following examples:
  - The cause of meat being halal is the slaughtering.
  - The cause of the Dhuhr salat being obligatory is the sun passing its zenith.
- ☐ Condition (*shart*) – a condition of zakat being obligatory is the passing of a lunar year, but it is not the cause, as further conditions must also be met (e.g. the *nisab*).
- ☐ Prevention (*mani'*) – the prevention of a woman's salat is menstrual bleeding.

CATEGORIES OF LEGAL RULINGS

50-52 Legal rulings of the *Shari'a* are five:

1. Obligatory – which is a strict demand to perform an action.
2. Recommended – which is a demand that is not strict to perform an action.
3. Disliked – which is a demand that is not strict to refrain from an action.
4. Prohibited – which is a strict demand to refrain from an action.
5. Permissible – which allows one to do or not do an action.

All of these are fully explained as follows:

Obligatory (*wajib* or *fard*): If the demand by the Lawgiver is strict it is called *wajib* or *fard* and it is rewarded if performed and punished if not, such as believing in Allah and His Messenger ﷺ and performing the five pillars.

Recommended (*mandub*): If the demand by the Lawgiver is of a lower level of strictness, it is categorised as *mandub*, which is rewarded if performed and not punished if not, such as praying Fajr (which is the two rak'as prayed before the *fard* Subh salat).

Disliked (*makruh*): If the prohibition is not strict it is called *makruh*, which is not punished if performed but rewarded if not, such as reciting Qur'an in sajda.

Forbidden (*haram*): If the prohibition is strict, it is called haram, which is punished if performed but rewarded if not, such as eating pork and drinking wine.

Permissible (*mubah*): If both performing an action and leaving it undone are permissible, it is called *mubah* or halal, which is neither rewarded nor punished. But if something halal is done with the intention of obeying Allah and avoiding disobedience, it is rewarded.

TYPES OF OBLIGATION AND RECOMMENDATION

Obligation (*fard* or *wajib*) divides into two: 53

1. *Fard 'ayn*, which is an obligation on every responsible individual (*mukallaf*), such as the five daily salats.
2. *Fard kifaya*, which is a communal obligation – such as *salat al-janaza* or saving a drowning person. If not performed by some of the community the entire community is responsible and liable for punishment, but if performed the community is discharged of the obligation.

The recommendation (*mandub*) divides into three:

1. Sunna is what the Messenger ﷺ always did without indicating it as an obligation. This is also called sunna *mu'akkada* (confirmed sunna) – and it comprises two types:
   - □ Sunna *'ayn*, the individual sunna – such as praying *salat al-witr.*
   - □ Sunna *kifaya*, the communal sunna – such as a member of group returning a greeting on the group's behalf, or calling the adhan or the *iqama*.
2. *Mustahabb* is what the Messenger ﷺ did sometimes.
3. *Tatawwu'* (voluntary) is any supererogatory act that a person performs of his own volition. This category is also called *nafila*, *raghiba* and *fadhila*.

The exact definitions and relationship between the many categories listed under 2. and 3. are topics of discussion between scholars.

Purification

Purification

INTRODUCTION

THIS SECTION ON the formal prayer (salat), the second pillar of Islam, starts with purification as it is a condition of the validity of salat.

PURITY

Water for Purification

Water is the instrument of purification unless unavailable. There are two types of water: 54-56

1. Unaltered. All water may be used for purification if unaltered in smell, taste and colour by something added to it, such as sugar, soap or objects falling into it, like food or earth.
2. Altered. Water has three rulings if altered by:
   - □ Impure substances, as described below. It must be thrown away.
   - □ Pure substances, as described below. It may be used for cooking, washing and such, but not for purification.
   - □ Unavoidably present natural substances. It may be used for purification. This includes:
     - Salt in seawater
     - Colour alteration caused by the surrounding earth
     - The green tinge from algae in a pond
     - A prolonged stay in a container
     - Melted ice and snow

Types of Purification

Purification is of two types:

1. Physically removing impure substances (*khabath*), which is discussed below
2. Removing ritual impurity caused by events (*hadath*), which is of three kinds:
   - □ Wudu'
   - □ Ghusl
   - □ Tayammum, which takes the place of wudu' and ghusl in special circumstances

Impure Substances

Impure substances (*najasa*) include the following:

1. Faeces from humans and animals that are forbidden to eat, such as pigs or dogs
2. Urine from humans and from animals that are forbidden to eat
3. Semen (*mani*) from men and animals
4. Pre-ejaculatory fluid (*madhy*)
5. Prostatic fluid (*wady*)
6. Blood and pus
7. Substances that come out from a dead body, like urine, saliva, tears, mucus, milk and eggs
8. Body parts supplied by blood severed from a dead, un-slaughtered or living animal, like skin, claws, fangs, teeth, flesh, bones, etc.
9. Altered and regurgitated food
10. Intoxicants (liquids)
11. Any part of a pig, including its milk

Impure substances must be removed from the body, clothes and the place of salat – meaning the actual ground touched by the forehead, hands, knees and feet.

They are removed by washing with water. If a stain or smell cannot be removed it is acceptable. If uncertain whether an impurity is present, sprinkling water purifies the doubtful area.

All ground is a place of salat for the *Ummah* of the Messenger ﷺ.

WUDU'

Obligations of Wudu'

There are seven obligatory elements of wudu': **57-61**

1. Rubbing as the water is poured or after it
2. Performing the wudu' without long interruptions[59]
3. Having the intention at the beginning to either:
   - □ Remove the state of ritual impurity or
   - □ Perform the obligation of wudu' or
   - □ Make permissible what was forbidden
4. Washing the face from the normal hairline to the chin and from ear to ear. If skin is seen through the beard the water must be made to reach the skin by combing with the fingers, and, if not, the entire beard must be wiped and the same goes for the eyebrows.
5. Washing the hands and arms, up to and including the elbows. Using the fingers to rub between the fingers of the other hand is obligatory.
6. Wiping the head, from the normal hairline to the base of the skull, and in the case of long hair, wiping to its extent.
7. Washing both feet up to and including the anklebones.

Sunan of Wudu'

There are also seven *sunan* of wudu': **62-63**

1. Washing each hand separately, before putting them in the water pot
2. Wiping back from the base of the skull towards the forehead, ending where you started with the obligatory
3. Wiping the ears once with wet forefingers and thumbs
4. Rinsing the mouth by moving water from side to side
5. Sniffing water up the nose (be careful if fasting that it does not reach the throat)
6. Blowing water out of the nose (avoid spraying by squeezing lightly the nostrils with the left hand)
7. Following the correct order of the obligatory elements[60]

---

59 See commentary of verse 69.

60 See previous section.

Recommended Elements of Wudu'

64-67 There are eleven recommended elements of wudu':

1. Saying the *basmala* ("*bismillah*" – "In the Name of Allah") before beginning
2. Choosing a clean place, so the water does not splash back impurities
3. Using little water
4. Placing an open pot on the right or a pitcher on the left[61]
5. Washing two or three times, even if the first time covered the body part totally
6. Beginning with the right limb before the left
7. Cleaning the teeth with the forefinger and thumb or *miswak*[62]
8. Correct order of the sunna
9. Correct order of the obligatory and sunna combined
10. Beginning from the forehead when wiping the head
11. Rubbing between the toes

Disliked Actions in Wudu'

68 There are two disliked (*makruh)* elements of **wudu'**:

1. Exceeding the amount of times prescribed when wiping the head and ears as described
2. Exceeding the amount of times prescribed when washing the body parts as described – except the feet; there is a difference of opinion whether they are only washed thrice, or may be washed until they are clean.

Cases in Wudu'

69 The definition of a short interruption is the time period in which a washed body part does not dry in moderate weather. A long interruption is what exceeds this.

61 Using the tap is acceptable providing we do not waste a lot water. And Allah knows best.

62 A teeth cleaning twig.

Interruptions in wudu' have the following rulings:

1. If one interrupts wudu' due to an inability to continue (e.g. one has spilt the water and must go to get more) it is continued from where it was interrupted, if the interruption was short (see above). If longer, one must start again.
2. If one interrupts wudu' due to forgetfulness, it is continued from where it was interrupted regardless of duration.
3. But if one interrupts it intentionally above the limit, starting again is necessary.

If forgetting an obligation of wudu', there are two rulings: **70-71**

1. If a long time has passed since the wudu' was finished, redo the obligation by itself.
2. If only a short time has passed before one remembers, redo wudu' from (and including) the missed obligatory element.

The rulings for missed actions in wudu' are:

1. A missed obligation invalidates the salat.
2. A forgotten sunna does not invalidate the salat, although this sunna should be performed by itself to be included for the next salat.

If one intentionally drops an obligation, and one changes one's mind and wants to perform it, the ruling depends on the amount of time which passed:

1. If a long time has passed, wudu' must be started again.
2. If a short time has passed one performs the missing element and everything after it.

If one intentionally drops a sunna and prays, it is *mustahabb* to do all the sunna and redo the salat if it is still within its time.

Things that Break Wudu'

Things which break wudu' are in two categories: **72**

1. Accidents (*ahdath*) – which break wudu' by themselves, such as urinating, passing wind, defecating
2. Causes (*asbab*) – which indirectly break wudu', such as sleeping,

which may provoke passing wind, or touching the opposite sex, which may provoke pre-ejaculatory discharge

There are sixteen accidents and causes that break wudu':

1. Urinating
2. Passing wind from the anus
3. Incessant flow or incontinence (*salas*) of:
    - □ Urine
    - □ Gas
    - □ Pre-ejaculatory discharge
    - □ Menstruation when over the set limit (*istihada*)[63]

    If these last four occur half the time or more from noon until sunrise, wudu' is not broken, although redoing it is recommended, but if they occur less than half the time, it breaks wudu'.

73 4. Defecation
5. Long or short deep sleep (i.e. being oblivious if one's name is called). A long light sleep does not demand wudu' (but it is recommended). A short light sleep is ignored
6. Pre-ejaculatory discharge (*madhy*), which is a thin transparent liquid caused by foreplay or fantasising, makes it obligatory to wash the full length of the penis before wudu'
7. Intoxication regardless of the means
8. Unconsciousness
9. Bouts of insanity, including epileptic fits
10. Emitting prostatic fluid (*wady*), a thick white discharge after urination

74-75 11. Touching or being touched (body, hand or lips, skin to skin or through clothing) if:
    - □ Pleasure was sought, even if it was not actually found, or
    - □ Pleasure was found, even if it was not sought – unless pleasure was not sought and not found, in this case wudu' is not broken
12. Kissing skin is as above, but kissing the lips breaks wudu', regardless
13. Inserting the fingers between one's labia, but touching the outside does not

63 If incontinence is easily treated, it breaks wudu', regardless of its frequency. If it occurs continuously it is not even recommended to do wudu'.

14. Touching one's penis with the inside or side of one's hand or fingers, unless there is a barrier, even if it is thin
15. Uncertainty (*shakk),* whether one has broken wudu' or not, breaks wudu' unless that thought occurs at least once every day. In that case the doubt is disregarded
16. Apostasy – may Allah protect us from it!

And if any of these occur while doing wudu' or after finishing, it must be redone.

CLEANING

Cleaning the Body of Impurities

When defecating or urinating it is obligatory to relieve the need completely. **76-77**
Empty the urethra by applying pressure by pulling the penis with the thumb and index finger of the left hand from the root to the head – lightly, to avoid incontinence in the long run.

There are three methods of cleaning oneself afterwards, listed in order of least preference:

1. Using stones (*istijmar*) – and also pieces of toilet paper – is permitted in order to clean urine and faeces, if the last one used is completely clean.
2. Using water (*istinja'*) is superior, and must be used if the impurity is much or spread about.
3. A combination of these, which is best.

However, women must always use water after urinating.

GHUSL

Obligations of Ghusl

There are four *obligations* of **ghusl**: **78-80**

1. Having the intention to either:
   - □ Perform the obligation of ghusl
   - □ Make permissible what was forbidden

- ☐ Eliminate ritual impurity

2. Doing it without long breaks.[64] If the break is long, three situations apply:
   - ☐ If the break is intentional, one must start over.
   - ☐ If due to forgetfulness, the ghusl is continued where left off.
   - ☐ If due to inability, (e.g. one has spilt the water and must go and get more) one must start again.
3. Washing the entire body by rubbing with the hand (or an instrument, if one cannot reach)
4. Combing through the hair with wet fingers, including beard and eyebrows. (Plaits are ignored, unless too tight for water to reach the scalp)

Make sure you get to all hidden places, such as the back of the knees, armpits, navel, inguinal folds, between the buttocks, the soles of the feet and between fingers and toes.

Wipe difficult spots, such as the middle of the back, with a long-handled brush, towel or get someone to rub it for you [another opinion in the school is that pouring water on such spots suffices].

Sunan of Ghusl

81 There are four *sunan* of ghusl:

1. Rinsing the mouth once
2. Beginning by washing the hands to the wrists once
3. Sniffing water up the nose and blowing it out once and
4. Wiping the auditory meatus with a wet finger (washing the ears themselves is *wajib*)

Recommended Elements of Ghusl

82-83 There are seven recommended elements of ghusl:

1. Start by removing impurities after washing the hands as described above.

64 See commentary verse 69.

2. Say the *basmala* (*bismillahi-r-rahmani-r-rahim*) before beginning.
3. Pour water over the head three times (this is the only action done thrice in ghusl).
4. First do the limbs that are normally washed in wudu' (but intending ghusl).
5. Use little water.
6. Start from the top of the body.
7. Start from the right side of the body.

Wash impurities from the private parts and other parts of the body and after that avoid touching your genitals with your palm, the side of the hand or fingers, so your *ghusl* stands for your wudu'. **84-85**

But if the genitals are touched with the palm or the side of the hand or the inner side of the fingers during or after the wudu, the ghusl is not annulled, but wudu must be redone.

Description of Ghusl

The above may be combined in eight sequential elements in *ghusl*:

1. Make the intention.
2. Say the *basmala*.
3. Wash the hands to the wrists.
4. Wash impurities from the private parts and other parts of the body.
5. First wash the limbs of wudu' once, and continue wudu' as described.
6. Comb through the hair with wet fingers.
7. Pour water over the head three times, washing the head, including the ears.
8. Divide the body into four parts washing it in the following order:
   - □ Top right
   - □ Top left
   - □ Bottom right
   - □ Bottom left

Things that Make Ghusl Obligatory

Ghusl is obligatory after: **86**

1. The termination of menstruation
2. The termination of (post-natal) bleeding
3. Orgasm with simultaneous ejaculation[65] achieved under the usual circumstances, which means the following cases do not require ghusl:
   - □ Orgasm without ejaculation
   - □ Ejaculation without orgasm (but in case of ejaculation with delayed orgasm, one must do ghusl according to the strongest opinion)
   - □ Ejaculation caused by unusual circumstances, such as:
     - Riding an animal
     - Being stung by a scorpion or bitten by a snake
     - Scratching oneself due to sickness such as scabs, etc.
4. The disappearance of the head of the penis into the vagina or anus, with or without ejaculation or erection – for both parties regardless of gender or age, whether dead or alive, human or not
5. Death (see the section on *salat al-janaza*)
6. When a disbeliever becomes Muslim

87-88 Until *ghusl* is performed, nos. 1 and 2 (menstruation and postnatal bleeding) prohibit three things:

1. Intercourse
2. Touching a *Mushaf* (a written copy of the *Qur'an*) except if studying or teaching from it. [However one may recite from memory or computer display, etc.]
3. Entering the salat area of a mosque

Until ghusl is performed, the state of major ritual impurity (*janaba*) caused by orgasm or penetration prohibits three things:

1. Recitation of Qur'an
2. Touching a *Mushaf*
3. Entering the salat area of a mosque

But repeated intercourse is allowed.

65 Or, in the case of a woman, increased effusion of fluid from the vagina.

The rulings for forgetting elements of ghusl are similar to those for wudu' except that one need never perform the parts after the one missed – i.e. just perform the missing part by itself. So if a person forgets to wash a spot of his body, he should just wash that part as soon as he remembers without it being necessary to repeat the washing of what comes after that spot. If a person remembers after having prayed, he should wash only that part and repeat the salat or salats made before having remembered.

TAYAMMUM

Reasons for Tayammum

Tayammum is purification without water which replaces ghusl and wudu' in two cases: **89**

1. If one fears harm:
   - ☐ To one's health by
     - Possibly causing an illness (such as ghusl in very cold temperatures)
     - Delaying the healing of a wound or recovery from an illness
     - Aggravating a current wound or condition
   - ☐ From enemy soldiers, thieves or beasts which block access to water
   - ☐ If hampered by physical inability, such as old age or illness
   - ☐ Risking thirst for oneself or one's animals
   - ☐ Depleting one's wealth because water is overpriced
   - ☐ If the majority of the body of the person doing ghusl or the majority of the limbs for the person doing wudu' are covered in wounds
2. If one lacks water, i.e.
   - ☐ If there is insufficient water
   - ☐ If there is nobody to bring it in the case of a disabled person
   - ☐ If there are no tools to retrieve the water, such as a well without a bucket
   - ☐ If it is the only chance to do the first *rak'a* of a *wajib* salat before its time ends

One should not search if one knows for certain no water is available. If finding water is plausible or uncertain[66] one must search for it, and if none is found, *tayammum* should be done. The more certain one is, and the more strength one has, the more effort must be put into the search, but it must not be a hardship, no matter who you are.

What Tayammum Permits

90-91 A single tayammum permits praying:

1. Only one obligatory salat and
2. Any consecutive and connected sunna salats to it
3. *Janaza* salat (salat for the dead)
4. Sunna salats or lesser ranked salats (such as *shaf'* and witr after 'Isha)
5. It also permits praying supererogatory salats by themselves on two conditions:
   - ☐ One is ill.
   - ☐ One is travelling (i.e. the healthy and/or resident person can only pray supererogatory salats using tayammum if they follow an obligatory salat as mentioned above).

Tayammum does not permit:

1. Praying an obligatory salat after a supererogatory salat, nor after using tayammum to touch a *mushaf* (in both cases tayammum must be repeated before the *wajib* salat)
2. Praying the Friday salat, unless one is travelling or ill (i.e. a resident group of people without water may only pray Dhuhr)
3. Praying supererogatory salats by themselves – unless one is travelling or ill, as mentioned above
4. If both water and pure earth are unavailable, then in this situation

---

66 The degrees of knowledge defined in the Shari'a are five:
Certainty (*yaqin*) – e.g. when many trustworthy people confer the knowledge.
Plausibility (*dhann*) – such as when a single reliable person tells it.
Uncertainty (*shakk*) – one considers the chances 50/50.
Implausibility (*wahm*) – when the matter in question is unlikely.
Ignorance, which is in two categories:
No knowledge at all (*jahl*) – i.e. one simply does not know, and is aware of one's ignorance.
Compound ignorance (*jahl murakkab*) – one believes one knows, and is unaware of one's ignorance.

the person is exempt from the *wajib* salat and this need not be done or made up.

Obligations of Tayammum

There are eight obligations of tayammum: 92-94

1. Wiping the face (but unlike in wudu' one ignores wrinkles, creases, and such)
2. Wiping both hands to the wrists, removing any rings and wiping between the fingers
3. Having the intention at the first touch of pure earth (*sa'id*) to make the salat permissible
4. First touch (written as *darba*, literally "strike," in Arabic) of the *sa'id*
5. Doing it without long breaks[67]
6. Pure earth (*sa'id*), i.e. the surface of the earth, regardless of whether it is sand, rock, mud or soil
7. Doing tayammum immediately prior to the salat, preferably before the *iqama*
8. Doing it after the arrival of the time of the salat, for which the recommended (*mustahabb*) times of making tayammum are:
   - □ At the end of the salat's *ikhtiyari*[68] time if hoping to obtain water
   - □ At the start of the salat's *ikhtiyari* time if despairing to obtain water
   - □ In the middle of the salat's time if one has uncertainty about obtaining water

Sunan of Tayammum

There are three *sunan* of tayammum: 95

1. Wiping the forearms from the wrists to the elbows
2. The second touching of the earth before wiping the arms
3. The order:

---

67 See commentary verse 69.

68 The *ikhtiyari* time is the period in which one has a choice when to pray (see *The Times of the Prayer*).

- □ First touch of the *sa'id*
- □ Wiping the face
- □ Second touch of the *sa'id*
- □ Wiping the arms to the elbow

Recommended Elements of Tayammum

96 There are two recommended elements of tayammum:

1. Saying the *tasmiya* or *basmala* (*bismillahi-r-rahmani-r-rahim*) before beginning
2. Doing it the praiseworthy way it has been transmitted from the Messenger ﷺ which is to:
   - □ Put the palm of the left hand on the back of the right hand.
   - □ Wipe upwards up to the elbow.
   - □ Turn the left hand and grab the arm with the thumb.
   - □ Turn over the right arm, facing the palm upwards.
   - □ Wipe downwards on the inside of the arm to the end of fingertips with the left hand – remembering that wiping between the fingers is obligatory.

What Breaks Tayammum

96-98 Tayammum is annulled by two things:

1. Whatever breaks wudu'
2. If water becomes available. However, the rulings depend on when it happens:
   - □ If water becomes available before the salat, tayammum is nullified – except if doing wudu' (or ghusl) would prevent the performance of the salat inside its time).
   - □ If water becomes available during the salat, one keeps praying (but if one remembers while praying that one actually had water, tayammum is nullified).
   - □ If water becomes available after the salat, the salat remains valid, but it is recommended to repeat it inside its *ikhtiyari* time in the following cases:
     - If one's access to water was blocked by thieves or wild animals

- If one hoped to find water, but prayed anyway
- In the case of the bed-ridden person who had no one to fetch water for them.

KHUFFS

When wearing *khuffs* (leather socks) one may dispense with washing the feet during wudu' by wiping the entire *khuff* with a moist hand whilst taking into consideration the following points:

1. The *khuffs* must be made of leather from a halal animal.
2. They should be stitched, not glued.
3. They should cover the area of the foot to be washed in wudu'.
4. They should be intact with the following two exceptions:
   - □ The length of a tear must be less than a third of the length of the foot.
   - □ A hole must be so small one does not feel water penetrating when wiping.
5. They must be put on and taken off as a pair.
6. They must be put on in a state of wudu'.
7. When wiping them, the entire surface of the *khuffs* must be accessible to water – remove socks, dried mud and the like.
8. Non-leather socks may go under the *khuffs*.
9. An unbroken wudu' is maintained upon removal of the *khuffs*. If the wudu' was broken and then redone by wiping the *khuffs*, in order to keep the wudu' both feet must be washed shortly[69] after the *khuffs* are removed, otherwise the wudu' is broken.

69 The definition of a short interruption is the time period where a washed body part does not dry in moderate weather.

The Salat

The Salat

ELEMENTS OF THE SALAT

Preconditions of the Salat

THE WORD *salat* (prayer) comes from *silah* and there are two opinions 99
about its linguistic roots:

1. That it means 'connection,' referring to between man and his Lord.
2. Or the original pre-Islamic meaning of *du'a* (supplication).

The salat has sixteen obligatory elements and some external conditions, which are of two kinds:

1. Four conditions to be fulfilled (*shart ada'*), because they are under the *mukallaf's*[70] control, are described in verse 105.
2. Six conditions of obligation (*shart wujub*), because they are not under the *mukallaf's* control, are described in verse 110.

Obligations of the Salat

There are sixteen obligations of salat, listed in order of performance: 100

1. The initial "*Allahu akbar*" (*takbirat al-ihram*) by which one enters a state of sacredness excluding all other actions. It must be short from the imam.[71] It is recommended to always say it aloud, and not before the lines have been straightened.
2. Standing (*qiyam*) during the *takbirat al-ihram* (except for the latecomer to the salat, who may initiate the *takbir* standing, and

70 Responsible person.

71 It is a sign of an imam's lack of knowledge of the fiqh that by lengthening the *takbirat al-ihram* he may cause the one following him to finish uttering his *takbirat al-ihram* before the imam's thereby invalidating the salat of the follower.

finish it in the bowing or prostrating position).[72] [73]

3. Having the intention during it, i.e. knowing which salat one is performing[74]
101 4. Reciting Surat al-Fatiha immediately after the *takbir*.[75] The commentator Mayyara mentions four points:
    - □ The Fatiha is not obligatory when praying behind an imam.
    - □ There is no du'a between the *takbirat al-ihram* and the Fatiha.
    - □ There is no seeking refuge from Shaytan (*ta'awwudh*) nor *basmala* – except in non-obligatory salats.
    - □ It is obligatory to learn al-Fatiha by heart, so until one learns one must pray behind an imam but if there is no teacher and no imam, reciting it is not obligatory.
5. To remain standing (*qiyam*) during its recitation.
6. Bowing (ruku') – the minimum being to bow so the palms are close to the knees. It is recommended to:
    - □ Put the palms on the knees.
    - □ Keep the knees straight.
    - □ Separate the elbows from the body by bending them slightly (for men only).
    - □ Keep the back and neck level.
7. Standing up after bowing – leaving it[76] invalidates the rak'a.[77]
8. Prostrating (*sajda*) with humility (*khushu'*), which includes the recommendation of putting the face and hands directly on the ground, unless it is very hot or cold. Prostration is on these points:
    - □ Hands – which should touch the ground first when going into sajda after standing
    - □ Knees – which should leave the ground first when standing up

---

72 There is another position that says that the latecomer has to utter the *takbirat al-ihram* whilst standing up.

73 This is only necessary in obligatory salats, because a non-obligatory salat can be performed whilst seated.

74 If one finds he intended a salat already done, e.g. intending Dhuhr at 'Asr, he should abandon it and start again with the correct intention for 'Asr.

75 But when the imam is reciting aloud, always be quiet and listen.

76 Some people drop down into prostration before standing up properly and thereby neglect this obligation.

77 A rak'a is a unit of the salat consisting of standing, bowing, rising from bowing, prostrating, sitting and prostrating. Prayers are made of a number of rak'as (one, two, three or four).

- ☐ Face – meaning the forehead and nose – which has two sub-rulings:
  - If the nose does not touch the ground, it is recommended to repeat the salat in its time.
  - If the forehead does not touch the ground, the salat is invalid and must be repeated.
- ☐ Toes – which should remain pointing towards the direction of the qibla

9. Rising from prostration. 102
10. Saying *"as-salamu 'alaykum"* (*taslim*) once (not *"salam alaykum"*, *"salamun 'alaykum"* or *"salamu 'alaykum"*) to end the salat. The imam should not lengthen its utterance.
11. Sitting (*julus*) for enough time to utter the salam. Sitting longer is sunna. It is recommended to sit putting the left buttock on the floor and the right shin over the left foot, while putting the bottom part of the right big toe or its side on the floor (this position is called *tawarruk*), placing the palms on the thighs.
12. Performing the obligatory elements in their correct order – i.e. standing before bowing, bowing before prostration and prostration before sitting. (Note that the order of the obligatory and sunna elements and the order of the sunna elements themselves is sunna).
13. Being straight and balanced when standing. 103-4
14. Resting briefly in each position, i.e. stillness of the limbs.
15. The follower (*ma'mum*) should never utter the *takbirat al-ihram* or the salam before or with the imam, as this invalidates his salat. To precede the imam in anything else is forbidden, but does not invalidate the salat; and to do anything else at the same time as the imam is disliked.
16. The follower must intend the same salat as the imam and follow him in it. The imam does not need to intend leading except for:
    - ☐ The fear salat
    - ☐ When joining Maghrib and 'Isha because of rain
    - ☐ The Friday (Jumu'a) salat
    - ☐ When replacing an imam who had to leave during the salat

And humility (*khushu'*), which is alluded to in point 8, is counted by some 'ulama as obligatory.

Conditions to be Fulfilled

105 Of the ten conditions of the salat, the conditions of performance (*shart ada'*), are demanded because they are under the *mukallaf's* control and there are four:

1. Facing the qibla from beginning to end if one remembers and is able. If neglected intentionally, the salat is invalid, except supererogatory salats on a mount.[78] If neglected by mistake or forgetfulness, it is recommended to repeat the salat within its time.
2. Removing physical impurities (*khabath*) from clothes, place and body, from beginning to end if one remembers and is able (failing this, it is recommended to repeat the salat in its time) but if one intentionally prays with filth that can be removed, the salat is invalid.
3. Covering nakedness (*'awrah*) from the beginning to end, if one remembers and is able. There are two types of nakedness, light and gross:
   - □ For men, light nakedness is everything between the navel and knees; and gross nakedness is the genitals and the anus.
   - □ For women, light nakedness is everything except the hands and face; and gross nakedness is everything from below the breasts to the knees.

   Intentionally exposing light nakedness is forbidden, but the salat is valid and it is recommended to repeat it inside the time. Exposing gross nakedness invalidates the salat.
4. Being free of ritual impurity (*hadath*) from beginning to end, regardless.[79]

106-7 The first three of these are necessary only when remembered and when one is able, but the fourth condition – being ritually purified – must be fulfilled in all cases.

There are many other rulings related to the one who forgets or is incapable:

1. For the one who forgot the first three it is recommended to repeat the salat in its time.

78 This also applies to buses, trains, etc. One simply indicates *ruku'* and sajda as best one is able. And Allah knows best.

79 See sections on *wudu'* and *ghusl*.

2. The one who is unable to face the qibla or cover the *'awra* is excused from doing these. However the one who calculated the qibla incorrectly, should repeat the salat within the time.

Nakedness that Women Should Cover

It is obligatory for a woman to cover everything except face and hands. During the salat, if chest, hair or limbs are revealed intentionally, it is recommended to repeat the salat within in its time.[80] **108-9**

NOTE: Exposing one's *nakedness* to anyone is haram, except one's spouse (or a doctor – if it is necessary). Catching a glimpse of one's own gross nakedness during salat invalidates the salat.

Conditions that Make the Salat Obligatory

The conditions of obligation (*shart wujub*), which are out of the *mukallaf's* control, are: **110-1**

1. Islam
2. Puberty
3. Sanity
4. That menstruation or post-natal bleeding have ended. The two signs of this are:
   - □ A cloth placed in a woman's private parts comes out wet with a white or transparent discharge or
   - □ A cloth placed in a woman's private parts comes out dry. Prayers missed during menstruation or post-natal bleeding need not be made up.
5. That the time of the specific salat has started. It is obligatory to perform the salat within its prescribed time.
6. And Qadi 'Iyad adds the arrival of *da'wa* for the non-Muslim.[81]

---

80 The time taken into consideration here is the *daruri* time, not the *ikhtiyari (see The Times of the Prayer).*

81 The author only mentions condition 4 and 5, but Mayyara mentions in his commentary the rest of the conditions under verse 99.

The Times of the Salat

With the exception of the obligatory morning salat (*Subh*),[82] the time of each of the obligatory salats divides into two:

1. The first period (*waqt ikhtiyari*) in which one has choice when to pray. If one discovers that he has prayed before this period, it must be redone immediately. It is forbidden to delay it beyond the *ikhtiyari* time without a valid pretext.[83]
2. The second period (*waqt daruri*) in which one has no choice and must pray, as soon as this time-period enters and the other conditions are fulfilled (such as making wudu', covering the *'awra*, removing impurities and facing the qibla).

   The period for *Subh* is treated as a first period (*waqt ikhtiyari*) and lasts from true dawn (*Fajr as-sadiq*), which is not the first light in the sky, but rather the thin line of light along the horizon that follows it, until the moment the disc of the sun emerges above the horizon (*shuruq*)

The two periods for Dhuhr are:

1. When the sun reaches the zenith until 'Asr begins
2. And from then until Maghrib gets close.[84]

The two periods for 'Asr are:

1. When an object's shadow equals its height, plus the length of its shadow at the zenith and it continues until the sunlight is reflected on walls and turns yellowish (*isfirar*),
2. And from then until Maghrib approaches.[85] (NOTE: the time of Dhuhr overlaps 'Asr)

---

82 There is an opinion in which the time of Subh also divides into *ikhtiyari* and *daruri*, the first being from Fajr (dawn) to *isfar* (when the light in the sky is such that the features of a person's face could be distinguished), and the second being from *isfar* to *shuruq* (sunrise). Both opinions are considered strong in the madhhab. I have chosen to mention the opinion that the time of Subh is not divided into *ikhtiyari* and *daruri* for the sake of ease especially for those living in the West and new Muslims, and Allah knows best.

83 Like being asleep, losing consciousness, etc.

84 I say "gets close" because time is needed to complete at least Dhuhr and one rak'a of 'Asr.

85 I again say "gets close" because time is needed to complete at least one rak'a of 'Asr.

The two periods for Maghrib[86] are:

1. When the disc of the sun disappears behind the horizon (sunset or *ghurub*) until the conditions of praying are fulfilled
2. And from then until Fajr comes close[87]

The two periods for 'Isha are:

1. When the red and yellow of twilight disappear (*shafaq*) from the sky (although light may remain) until the first third of the night has passed (NOTE: the night starts at Maghrib)
2. And from then until Fajr comes close[88] (NOTE: the time of Maghrib overlaps 'Isha)

In a mosque, the adhan and the salat must always be at the beginning of the time.

Essentially, the salat times rely on observation. If the signs are invisible, such as in cities with excessive illumination at night which prevent seeing the dawn or in cloudy weather, it is acceptable to follow astronomical calculation of the times. However, the times of *al-Fajr as-sadiq* and the beginning of 'Isha can never be accurately calculated.[89]

Confirmed Sunan of the Salat

There are twenty-two *sunan* of salat in total. The first ten are confirmed sunan (*mu'akkada*) for the person praying alone and the imam. Failing to perform them demands prostrations of inattentiveness: **112-5**

1. Reciting a *surah*, or even an *ayah*, after the Fatiha in the first two

86 This is the strongest opinion, i.e. Maghrib has no *ikhtiyari* period and must be prayed as soon as its conditions are fulfilled, but that the *daruri* period lasts until *al-Fajr as-sadiq*. Another opinion, which we cite for the ease of Muslims in the West, is that its first period is from sunset until 'Isha and the second from 'Isha until *al-Fajr as-sadiq*.

87 I say "comes close" because time is needed to complete at least Maghrib and one rak'a of 'Isha.

88 I again say "comes close" because time is needed to complete at least one rak'a of 'Isha.

89 I have come to this conclusion after realising that the use of angles (18°, 17°, 15°, etc) are not very accurate when compared with well established observation, especially when dealing with latitudes over 50°, and Allah knows best.

rak'as for the one who prays alone and the imam, and for the follower it is recommended to listen to the recitation of the imam (The recitation of the surah after the Fatiha is only considered recommended in supererogatory salats)

2. Standing during the recitation
3. Reciting aloud in the first two rak'as of Maghrib, 'Isha, Subh and Jumu'a. This has two rulings depending on gender:
   - □ For men:
     - The minimum volume is that the reciter and whoever is next to him can hear the recitation.
     - The maximum volume has no limit.
   - □ For a woman, the maximum volume is that the reciter can hear herself, but not the person next to her.
4. Reciting silently when praying Dhuhr, 'Asr, the last rak'a of Maghrib and the last two rak'as of 'Isha, which also has two rulings depending on gender:
   - □ For men:
     - The minimum is moving the tongue without sound.
     - The maximum volume is that he can hear himself.[90]
   - □ Women only move the tongue.
5. Saying all other *takbirs* apart from the *takbirat al-ihram* (which is obligatory)
6. Saying the first *tashahhud*
7. Saying the second *tashahhud*
8. Sitting during the first *tashahhud*
9. Sitting during the second *tashahhud*[91]
10. Rising from ruku' while saying *sami'a Allahu liman hamidah* (Allah hears the one who praises Him). The answer of the follower saying *rabbana wa laka al-hamd* (Our Lord, and to You is the praise) is just recommended.

90 When two or more late-comers stand up to complete their salats and the missed rak'a had a loud recitation, they should switch to silent recitation in order not to disturb one another, because to recite aloud in its proper place is sunna, but to disturb a person in the salat is forbidden, and the avoidance of the forbidden takes precedence over the obligatory, let alone doing what is only sunna.

91 Sitting while making the salam is obligatory.

Sunan of the Salat

The above ten are sunna *mu'akkada* (confirmed sunna) for the person praying alone and the imam and failing to perform them demands prostrations of inattentiveness (*sajda as-sahw*). The remaining twelve *sunan* listed below are treated as recommended, in that leaving them undone does not demand prostrations of inattentiveness. 116-7

11. The call to commence (*iqama*), between the adhan and the salat, even if making up salats. For men, it is aloud for every obligatory salat, even if praying alone. It is recommended for women praying alone to do it silently. The words are: "*Allahu akbaru, Allahu akbaru, ashhadu alla ilaha illa Allahu, ashhadu anna Muhammadan rasulu llahi, hayya 'ala-s-salati, hayya 'ala-l-falahi, qad qamati-s-salatu, Allahu akbaru, Allahu akbaru, la ilaha illa Allah.*"[92]
12. During prostrating the following parts of the body should be touching the ground:
    - □ Hands adjacent to, but not further forward than the ears, pointed towards the qibla and forearms above the ground (men should keep their elbows away from their sides and women should keep them by their sides)
    - □ Toes remain pointed towards the qibla
    - □ Knees so that thighs and belly are separate for men but close for women
13. The follower is silent and listens to the imam when he recites aloud, whether he can hear him or not. 118-9
14. The follower returns the salam to the imam turning his head to the front (even if he joined to pray only one rak'a).
15. The follower returns the salam to the left only if anyone is praying there.
16. Pausing for more than the minimum stillness required in each position.
17. Setting up a *sutra* if the imam or person praying alone believes people may pass through the area of prostration. The *sutra* is any object or creature, but must be:

92 Its meaning being: *Allah is the Greatest, Allah is the Greatest, I testify there is no god but Allah, I testify Muhammad is the Messenger of Allah, come to salat, come to success, the salat has been established, Allah is the Greatest, Allah is the Greatest, there is no god but Allah.*

- □ Ritually clean
- □ Immobile
- □ Have the minimum thickness of a spear
- □ Be at least a cubit long (length of the forearm)
- □ And should not distract the person praying.

120-1 18. For everybody to say the first salam out loud to end the salat. All the other salams should be said silently. The imam must keep his salam short.[93]

19. That wording of the *tashahhud*, which is what Malik reported from 'Umar ibn al-Khattab in the *Muwatta'*: *At-tahiyyatu lillah, az-zakiyatu lillah, at-tayyibatu-s-salawatu llilah. As-salamu 'alaika ayyuha-n-nabiyyu wa rahmatu-llahi wa barakatuh. As-salamu 'alaina wa 'ala 'ibadi-llahi-s-salihin. Ashhadu alla ilaha illa-llahu wahdahu la sharika lah, wa ashhadu anna Muhammadan 'abduhu wa rasuluh.*[94]

20. The prayer upon the Messenger (*as-salatu 'ala n-nabiy*) after the last *tashahhud*. Its wording is: *Allahumma salli 'ala Muhammadin wa 'ala ali Muhammad, kama sallaita 'ala Ibrahima wa 'ala ali Ibrahim, wa barik 'ala Muhammadin wa 'ala ali Muhammad, kama barakta 'ala Ibrahima wa 'ala ali Ibrahim. Fi-l-'alamina innaka hamidun majid.*[95]

21. To call the adhan for the five salats if a group expects the general public to join; in open fields and where it is usual to gather for salat, such as mosques. The wording of the adhan is:

Allahu akbar[96]

Allahu akbar

93 If the imam lengthens the initial *takbirat al-ihram* or the final salam, he could make the follower start before him or finish before him and that will invalidate the follower's salat. Therefore the Imam should always shorten the initial *takbir* and final salam.

94 Greetings are for Allah, good actions are for Allah, good words and salats are for Allah. Peace be upon you, O Prophet, and the mercy of Allah and His blessings. Peace be upon us and upon the right-acting slaves of Allah. I testify there is no god but Allah, alone, without partner, and I testify that Muhammad is His servant and Messenger.

95 O Allah, bless Muhammad and the family of Muhammad as you blessed Ibrahim and the family of Ibrahim and pour *baraka* on Muhammad and on the family of Muhammad as you poured *baraka* on Ibrahim and the family of Ibrahim! In all the worlds. You are Praiseworthy, Glorious!

96 Allah is the Greatest.

Then the caller (*mu'adhdhin*) says in a low but audible voice:

Ash-hadu alla ilaha ill'Allah[97]

Ash-hadu alla ilaha ill'Allah

Ash-hadu anna Muhammadan rasulullah[98]

Ash-hadu anna Muhammadan rasulullah

Then the caller raises his voice once more, saying:

Ash-hadu alla ilaha ill'Allah

Ash-hadu alla ilaha ill'Allah

Ash-hadu anna Muhammadan rasulullah

Ash-hadu anna Muhammadan rasulullah

Hayya 'ala-s-salah[99]

Hayya 'ala-s-salah

Hayya 'ala-l-falah[100]

Hayya 'ala-l-falah

(Only in the Subh salat, add at this point:

As-salatu khayrun mina-n-nawm[101]

As-salatu khayrun mina-n-nawm)

(Then for every salat, continue:

Allahu akbar

Allahu akbar

---

97 I testify that there is no god but Allah.

98 I testify that Muhammad is the Messenger of Allah.

99 Come to the salat.

100 Come to success.

101 Salat is better than sleep.

La ilaha ill'Allah[102]

122-3 22. To shorten the salat if travelling, which makes permissible:
- ☐ Shortening Dhuhr, 'Asr and 'Isha down to their two first rak'as, once past the outskirts of one's hometown until returning to its outskirts on the way back
- ☐ To join two consecutive salats:
  - To pray Dhuhr and 'Asr in immediate succession at any time within the periods of Dhuhr and 'Asr
  - To pray Maghrib and 'Isha in immediate succession at any time within the periods of Maghrib and 'Isha

Shortening and joining prayers for travelling is allowed, on the following three conditions:

- ☐ The distance between the place of origin and one's destination is, or exceeds 4 *barids* which is the equivalent of 48 miles or 77.25 km.[103]
- ☐ One does not intend to stop in a place for a period of 20 obligatory salats or more. (If the duration of the stop is unknown, continue to shorten them until it is clear).
- ☐ The purpose of the trip is permitted, not forbidden.

Recommended Elements of the Salat

124-5 There are twenty-one recommended elements of the salat:

1. Turning one's head to the right for the *as-salamu 'alaykum* while pronouncing "*kum*"
2. Saying *"amin"* after *al-Fatiha*, except for the imam when he recites aloud but another opinion says he can. The *amin* has to be said silently because it is a supplication (*du'a*).
3. Saying "*Rabbanna wa laka-l-hamd*"[104] when getting up from ruku', except for the imam
4. The silent *qunut* supplication in the last rak'a of Subh, either after

102 There is no god but Allah.

103 In terms of time it is equivalent to a journey of a day and a night at the pace of normally loaded beast.

104 Our Lord, and to You is the praise.

the *surah*, which is better, or after rising from ruku'. It is as follows: *Allahumma inna nasta'inuka wa nastaghfiruka wa nu'minu bika wa natawakkalu 'alayk. Nakhna'u laka wa nakhla'u wa natruku man yakfuruk. Allahumma iyyaka na'budu wa laka nusalli wa nasjudu wa ilayka nas'a wa nahfidh. Narju rahmataka wa nakhafu 'adhabaka-l-jidd. Inna 'adhabaka bi-l-kafirina mulhiq.*"[105]

5. Wearing a loose long cloak, especially for the imam **126-8**
6. Praising Allah (*tasbih*) in both:
    - □ Bowing (*ruku'*) saying *"Subhana Rabbiya-l-'Adhim",*[106] or the like
    - □ Prostration (*sajda*) saying *"Subhana Rabbiya-l-A'la"*,[107] or the like
7. Leaving the hands by one's sides (*sadl*).[108] Folding the hands (*qabd*) is disliked in the obligatory salat.
8. Saying the *takbir* at the onset of movements with two exceptions:
    - □ When rising from bowing (when one says "*Rabbanna wa laka-l-hamd*" as mentioned)
    - □ When rising from the first sitting, when one does not say it until fully upright
9. The grip in *tashahhud*, which is to:
    - □ Clench the right little, ring and middle fingers
    - □ Point the thumb and the index finger straight towards the qibla
    - □ Let the side of the index finger face upwards

---

105 O Allah! Truly we seek Your help and Your forgiveness and believe in You. We submit ourselves to You and surrender, and we abandon all who disbelieve in You. O Allah! You alone we worship. We pray and prostrate to You. We strive and struggle in Your Way. We hope for Your mercy and fear Your harsh punishment. Certainly Your punishment encircles the disbelievers.

106 Praise be to Allah, The Vast.

107 Praise be to Allah, The High.

108 This is the *mashhur* (majority) and *rajih* (strongest) position in the *madhhab* based on Malik's opinion transmitted in the *Mudawwana*. Malik's proof for this is the practice of the people of Madina, which, for Malik, is stronger than the *hadiths* that transmit the *qabd*. The transmission of the *hadiths* of *qabd* by Malik in his *Muwatta'* does not contradict this, since it only proves that Malik was aware of the existence of the *hadiths*, yet he considers the *qabd makruh* in the obligatory salat because *sadl's* proof (*dalil*) was considered by Malik as stronger. As for the story of Malik being disabled and incapable of *qabd* due to a beating, it is nothing but a lie invented by those incapable of understanding the importance that Malik placed on the *'Amal Ahl al-Madina*.

   - □ Rest the bottom of the right clenched hand on one's right thigh above the knee

10. Moving the index finger during the *tashahhud*. There are different opinions about this:
    - □ Left-right
    - □ Up-down
    - □ The best is in a circular motion, which is a combination of left-right and up-down

129-34 11. That men keep thighs apart from the belly and elbows apart from the knees and their knees apart from each other during prostration and women do the opposites (keep them together).

12. The sitting position in *tashahhud* and between prostrations: keeping the left buttock and the bottom or outside of the right big toe on the ground and folding the left foot under the right shin. Sitting between the two sajdas is obligatory[109] but sitting for the *tashahhud* is sunna.
13. In ruku' to put one's hands on the knees
14. Keeping the knees straight
15. For the follower, to recite silently behind the imam who is reciting silently
16. Keeping the hands in line or no further forward than the ears in sajda
17. Raising the hands to chest or shoulder level for the *takbirat al-ihram* and bringing them down. Palms may face the qibla or the ground, which is better.
18. Selecting the length of the surahs (the imam of the *jama'a* bearing in mind the weak, busy or old):
    - □ Long for Subh and Dhuhr (Surah 49-79 / al-Hujurat to an-Nazi'at)
    - □ Medium for 'Isha (Surah 80-92 / 'Abasa to al-Layl)
    - □ Short for 'Asr and Maghrib (Surah 93-114 / ad-Duha to an-Nas)
19. Making the recitation of the first rak'a longer than in the second
20. For the two *tashahhuds*, to make the first sitting shorter than the second sitting
21. Touching the ground with the hands before the knees when going

109 Rising from sajda.

into sajda, and letting the hands be the last to leave the ground when rising

Disliked Actions in the Salat

There are fifteen disliked actions in the salat: **135-6**

1. Saying the *basmala* in the obligatory salat aloud or silently
2. Seeking refuge (*ta'awwudh*) with the words *"a'udhu billahi mina-sh-shaytani-r-rajim"* in the obligatory salat
3. Doing sajda on cloth or anything other than whatever ground there is, except a coarse straw mat, unless the ground is impure or very hot or cold. This ruling applies only to palms, forehead and nose.
4. And this includes doing sajda on:
   - □ One's turban. If it is thicker than two layers, the salat must be repeated.
   - □ One's sleeve
5. Carrying anything in one's sleeve
6. Having anything in one's mouth, because this will distract one from the salat
7. Reciting Qur'an in sajda and ruku'[110] **137-9**
8. To be thinking about worldly affairs – indeed anything that distracts from humility (*khushu'*) before the Lord of the Worlds (although it does not invalidate the salat)
9. Absent-minded fiddling with one's beard, ring, etc.
10. Looking around by turning one's head or body, but the salat is not invalidated, unless one actually turns one's back to the qibla
11. Making supplication (*du'a*) in the recitation or in ruku'[111]
12. Interlacing the fingers
13. Cracking the knuckles
14. Putting one's hands on hips or waist
15. Closing the eyes – except when seeing something distracting[112]

110 Except if it is a supplication from the Qur'an, then it can be recited in sajda.

111 In the obligatory salats, according to as-Sawi's *Bulghatu-s-Salik*.

112 Note that in the Maliki school, one should look straight ahead when praying.

TYPES OF SALAT

Classification of Salats

There are two types of salat:

1. Obligatory (*fard* or *wajib*), which subdivide into two categories:
   - □ The salat which is obligatory on every individual *mukallaf* (*fard 'ayn*)
   - □ The salat which is a communal obligation (*fard kifaya*) i.e. the funeral salat (*salatul janaza*)
2. Supererogatory (*nafila*): the rest which are not obligatory and subdivide into two categories:
   - □ The salats which have a specific name, such as:
     - The odd salat (*salat al-witr*): The single rak'a at the end of the day's salats
     - The solar eclipse salat (*salat al-kusuf*)
     - The lunar eclipse salat (*salat al-khusuf*)
     - The two festivals salat (*salat al-'idayn*)
     - The drought salat (*salat al-istisqa'*)
     - The dawn salat (*salat al-Fajr*): The two rak'as before the obligatory morning salat (*salat as-Subh*)
     - The salat to greet the mosque (*salat tahiyyat al-masjid*)
     - The midmorning salat (*salat ad-duha*)
     - The tarawih salat: The communal salat every night of Ramadan
   - □ The salats which have no specific name, such as the rak'as performed before and after the obligatory salats, or which are done at any time, except at times when praying is forbidden

The Five Prayers

140 The aforementioned obligatory salats, which are an individual obligation (*fard 'ayn*), are the five daily salats:

1. *Subh* – the morning salat of two rak'as
2. *Dhuhr* – the midday salat of four rak'as
3. *'Asr* – the afternoon salat of four rak'as
4. *Magrib* – the sunset salat of three rak'as

5. *'Isha* – the night salat of four rak'as

The Friday salat (*salat al-Jumu'a*) is the communal salat which replaces Dhuhr every Friday on certain conditions.[113]

These five daily salats are necessarily known as part of the Deen by all Muslims. Three scenarios apply regarding their obligatory nature:

1. A Muslim who denies the obligatory nature of the five prayers is an apostate (*murtadd*), i.e. has left Islam and, if he does not repent, he is to be executed by the legitimate Islamic authority for committing disbelief (*kufr*).
2. A Muslim who confirms the obligation of the salats, but refuses to pray them in the given time is to be executed by the legitimate Islamic authority for transgression (*hadd*), not for disbelief (*kufr*).
3. There is no punishment for someone who refuses to make up missed salat.

The aforementioned obligatory salat which is a communal obligation (*fard kifaya*) is the Funeral Salat (*salat al-janaza*) or the salat over the dead Muslim, if the body (or the larger part of it) is available for burial.

The Funeral Salat

There are six obligatory elements of the Funeral Salat: **141**

1. Four *takbirs*[114]
2. An unspecified supplication (*du'a*)[115] for the deceased after every

113 See section on conditions to be fulfilled in the Friday salat.

114 Hands are only lifted for the first.

115 The supplication chosen by Imam Malik in his *Muwatta'* and transmitted by Abu Hurayra is the following: *Allahumma innahu 'abduka wa-bnu 'abdika wa-bnu amatik. Kana yashhadu an la ilaha illa anta wahdaka la sharika laka wa anna muhammadan 'abduka wa rasuluka, wa anta a'lamu bih. Allahumma in kana muhsinan fazid fi ihsanihi, wa in kana musi'an fatajawaz 'an sayyi'atih. Allahumma la tahrimna ajrahu wa la taftinna ba'dah.* Which means: "O Allah! He is Your servant, son of Your male servant and Your female servant. He used to witness that there is no god but Allah without partner, and that Muhammad is Your Servant and Messenger. And You knew him best. Oh, Allah! If he was good increase his goodness, and if he was bad, overlook his bad actions. Oh, Allah! Do not deprive us of his reward and do not make us fall in temptation after him." If one just says: *Allahumma-ghfir lahu wa-rhamhu* "O Allah! Forgive him and have mercy

takbir, even after the fourth one[116]

3. Having the intention
4. Only one salam after the fourth *takbir*, the imam says it so those next to him can hear him and the followers say it silently
5. Standing throughout – no ruku' and no sajda
6. Doing it behind an imam

If possible, it is done at the burial site just before the body is lowered into the grave.

Death, Washing, Shrouding & Burying

142 When a Muslim is dying, that person should be with other Muslims who say the shahada until that dying person also says it, the objective being that the shahada are the last words uttered by that person before death. No *kuffar* or menstruating women should be present.[117]

It is *fard kifaya* to perform three actions for any dead Muslim:

1. A full washing (ghusl), similar to the ghusl for the state of major impurity, by:
    - ☐ Removing impurity. Any remains in the intestines are removed by lightly pressing on the stomach.
    - ☐ Beginning with the limbs of wudu'.
    - ☐ Washing the rest of the body, while it is covered by a cloth, under which one uses a thick piece of cloth to wash. Men wash men, women wash women.
2. Shrouding. Applying camphor or perfume to the seven points of prostration (*sajda*) is recommended. Bodies must be completely covered when buried:
    - ☐ For a dead man it is recommended to use three (or five) pieces of cloth:
        - A *qamis* (a long shirt, i.e. a *jalaba*)
        - A turban

---

on him" after each *takbir*, it will be enough.

116 The Fatiha is not recited.

117 This is stated in the *Risala* of al-Qayrawani, chapter *On what to do when someone is at the point of death.*

- An *izar* (a wrap-around from navel to feet), – but adding two extra big sheets totalling five items of clothing is better.

□ For a dead woman it is the same, but a headscarf (*khimar*) replaces the turban, and two extra wraps are added to make it seven.

3. Burial. The following points are noted in the *Risala* of al-Qayrawani:
   - □ It is recommended to participate in the funeral salat and the funeral procession.
   - □ Reciting *Sura Yasin* is recommended in connection with the funeral.
   - □ Crying is acceptable, unless overdone; screaming, tearing one's clothes, etc. are forbidden.
   - □ The deceased is buried lying on the right side with the chest towards the qibla.

A *shahid*[118] is an exception: he is not prayed over nor washed, shrouded or scented. Rather, he is buried in the clothes in which he was killed (but nakedness must be covered). Also, there is no funeral salat for the stillborn.

All of this is done according to the financial status of the deceased. The expenses related to the burial are paid by whoever is responsible for him, i.e. the father over the son, the son over the poor parents, the master over slave. But a wife's expenses are taken from her own estate. If the deceased has no money, it is taken from the Public Treasury (*bayt al-mal*). If there is none, the Muslim community should provide.

The Witr Salat

The *witr* is a confirmed sunna and the person who does not perform it is **142**
discredited as a trustworthy witness. It is distinguished from other salats by three aspects:

1. It is a single rak'a which should be prayed separately after *shaf'* (of minimum two rak'as).[119] It is recommended to recite sura no. 87[120]

118 A Muslim who died in battle whilst doing jihad.
119 Leaving the *shaf'* is *makruh*.
120 Al-A'la.

and 109[121] respectively in the two *shaf'* rak'as that precede the *witr.*

2. It is recommended to recite in the *witr* the last three suras[122] of the Qur'an after the *Fatiha.*
3. Its time is divided into two periods, similar to the obligatory salats:
   - □ *Ikhtiyari* starts when two conditions are fulfilled:
     - After the disappearance of the red twilight (*ash-shafaq al-ahmar*) from the sky
     - After *salat al-'Isha* has been prayed; it lasts until dawn (*Fajr*).
   - □ *Daruri*, which starts from dawn until *salat as-Subh* is prayed.

It is preferred that it is prayed at the end of the night, but whoever does not regularly wake up for night vigil salats (*tahajjud*) should pray it before sleeping.

The Solar Eclipse Salat

The salat of the solar eclipse (*salat ul-kusuf*) is a sunna. It is distinguished by these points:

1. It has no adhan or iqama.
2. It is prayed in congregation (*jama'a*) in the mosque, but may be prayed individually at home.
3. It is performed for the duration of the eclipse.
4. Its time is from when the supererogatory salat is allowed in the morning,[123] until the sun reaches its zenith.
5. The salat consists of two rak'as.
6. Each rak'a has two bowings (ruku') and two standings (*qiyam*).

According to what is in the *Risala* of al-Qayrawani, this salat is carried out in the following manner:

1. In the first standing, one recites silently for approximately the duration of *Surat al-Baqara.*
2. Then one bows.
3. Stands up with the words *"Sami'a-llahu liman hamidah"* (Allah hears the one who praises Him).

121 Al-Kafirun.

122 Al-Ikhlas, Al-Falaq & An-Nas.

123 That is when the sun is well clear of the horizon.

4. In the second standing, one recites again, this time for the approximate duration of Surah Al 'Imran.
5. Then one bows a second time.
6. Stands up with the words *"Sami'a-llahu liman hamidah"* into the third standing (without recitation).
7. Then one goes into sajda twice. The prostrations should be long.
8. In the fourth standing, one recites silently for approximately the duration of Surat an-Nisa'.
9. Then one bows a third time.
10. Then one stands up with the words *"Sami'a-llahu liman hamidah."*
11. In the fifth standing, one recites again, this time for the approximate duration of Surat al-Ma'ida.
12. Then one bows a fourth time.
13. Then one stands up with the words *"Sami'a-llahu liman hamidah,"* into the sixth standing (without recitation).
14. Then one goes into sajda twice, and it should be long.
15. Then one concludes with the *tashahhud* and salam in the normal way.

There is no inconvenience in giving a talk (*khutba*), but not as formal in content as the Jumu'a's talk.

The Lunar Eclipse Salat

The salat of the lunar eclipse (*salat al-khusuf*) is not a sunna, but *a recommendation* (*mustahabb*).

1. It is prayed individually at home. Doing it in groups is disliked.
2. It has repeated cycles of two rak'as until the eclipse ends.
3. The recitation is silent.
4. Its time limit is from sunset (Maghrib) to dawn (Fajr).

The 'Id Salat

The salat of the two *'Ids* is a confirmed sunna for every responsible person (*mukallaf*) who is:

1. Free, i.e. not a slave
2. Male
3. Resident, i.e. not travelling

If one cannot join a congregation (*jama'a*), one should pray it alone.

It is distinguished by ten aspects:

1. It has no adhan or iqama.
2. It has seven *takbirs* (i.e. the words *'Allahu akbar'*) in the first rak'a, including *takbirat al-ihram*.
3. It has six *takbirs* in the second rak'a, including the *takbir* of standing up.
4. Hands are raised only in the *takbirat al-ihram*.
5. The imam gives time for the followers to repeat the *takbirs*.
6. It is recited out loud.
7. It is recommended to recite *Surat al-A'la* in the first rak'a and *Surat ash-Shams* in the second.
8. Two *khutbas* are performed similar to the Jumu'a, but both are opened with *takbir*, which is repeated by the listeners.
9. It is done outside in an open area (*musalla*), unless conditions such as cold, rain, extreme heat or fear make this impossible, otherwise a mosque will suffice.
10. Its time limit is from when the supererogatory salat is permitted in the morning until the sun reaches its zenith on the day of 'Id.

The Salat for Rain

The salat for rain (*istisqa'*) is a sunna in periods of severe drought when water is needed for crops or drinking, whether the water comes from a river or rain. It is distinguished by the following aspects:

1. It is performed in an open area (*musalla*).
2. The congregation (*jama'a*) comes out on foot in humility, wearing worn clothes.
3. It is two rak'as.
4. It is recited out loud.
5. The imam gives two khutbas after it (similar to Jumu'a) which must include:
   - □ A lot of asking forgiveness (*istigfar*), instead of *takbir*
   - □ A long supplication (*du'a*) at the end of the second
6. Then, the imam turns to the qibla and turns his cape inside out.
7. The men behind him do the same while sitting.

8. It is recommended to fast three days, the third day being the one they set aside for the salat, and to give charity (*sadaqa*).

The Salat of Dawn

While all of the above, except the lunar eclipse, are sunna, the salat of dawn (Fajr) is a salat encouraged (*raghiba*) by both the word and practice of the Prophet ﷺ. He said: "The two rak'as of Fajr are better than the world and everything in it." (Muslim). It is distinguished by five aspects: **143**

1. No suras are recited after the Fatiha.
2. It is silent.
3. If there is very little time to pray Subh, Fajr is postponed.
4. Its time is from Fajr til the salat of Subh has been prayed. If missed, it can be made up between the sun becoming well clear of the horizon, until the sun reaches the zenith (*zawal*).
5. If one has not prayed Fajr and Subh before sunrise, one prays Subh first, then Fajr.

Note that other sunna salats are not made up if missed, unlike obligatory salats, which must always be made up as soon as possible and in their correct order, regardless of the time of day.

Prayers missed while travelling may be shortened when made up. Prayers missed due to sickness are prayed normally.

Other Supererogatory Prayers

Supererogatory (*nafila*) salats are always recommended, except in two cases: **144**

1. At times when praying them is forbidden:
   - ☐ During sunrise
   - ☐ During sunset
   - ☐ When the imam is on the *minbar*
2. At times when praying them is disliked (*makruh*):
   - ☐ After 'Asr
   - ☐ From sunset until the salat of Maghrib has been performed
   - ☐ From sunrise until the sun has come well clear of the horizon

Ibn Ashir mentions eight confirmed supererogatory salats: **145**

1. The salat to greet the mosque (*tahiyyat al-masjid*)
2. The midmorning salat *(Duha)*
3. The communal salat every night of the month of Ramadan (*tarawih*)
4. The two rak'as before witr (*shaf'*)
5. The two (or four) rak'as before Dhuhr
6. The two (or four) rak'as before 'Asr
7. The two (or four or six) rak'as after Maghrib
8. The two (or four) rak'as after Dhuhr

Unless otherwise indicated, recommended salats should be done:

1. In pairs of two rak'a, except witr
2. Silently during the day, out loud during the night
3. Not in group (jama'a), except tarawih

SALAT RECTIFICATION

Sajdas (Prostrations) of Inattentiveness

146 It is sunna to do two sajdas after the *tashahhud* and before the salam (*sujud qabli*) if one inadvertently misses out either:

1. A confirmed sunna (*mu'akadda*), such as reciting silently what is loud in the obligatory salat (see section on sunna *mu'akadda of salat*)
2. Two or more light *sunnas* (mentioned in the section on sunna of salat), such as missing two *takbirs* (except *takbirat al-ihram* which is obligatory) and two "*sami'Allahu liman hamidah*"

These sajdas are performed straight from the sitting position immediately after the *tashahhud* and before the salam. The person repeats the *tashahhud* after the two sajdas[124] and then utters the salam.

147 It is sunna to do two sajdas after the salam (*sujud ba'di*) if one inadvertently adds something, such as a fifth rak'a or recites loudly in a silent obligatory salat. These prostrations are performed straight from the sitting position[125]

124 According to the widespread opinion (*mashhur*).

125 Nevertheless, if an extra rak'a is added and the person remembers whilst standing during the extra rak'a, he should go straight away to perform the two sajdas of inattentiveness without starting them from the sitting position.

immediately after the normal salam. The *takbirat al-ihram* should be said before the two sajdas without lifting the hands. After the two sajdas, the person repeats the *tashahhud* and then utters the salam again.

Whoever both adds and subtracts, the subtraction takes precedence without exception, so one must perform the prostrations before the salam (*sujud qabli*).

Forgetting the Sajdas of Inattentiveness

If one forgets the prostrations of inattentiveness, the ruling depends on the type: **148**

1. If the *qabli* is forgotten, it must be done immediately after the salam. If it is not done immediately and more time passes than it normally takes to leave the mosque[126] after praying:
   - ☐ The salat is invalidated if the *qabli* was caused by missing three or more sunnas
   - ☐ The salat is still valid if the *qabli* was caused by missing less than three sunnas
2. If the *ba'di* is forgotten, the salat remains valid and the prostrations may be done even if years have passed

In both cases, one must finish it with a salam after *tashahhud*.

The imam assumes the responsibility of the mistakes of subtraction or addition by the follower (as long as he is following the imam) and this relieves the need for prostrations of inattentiveness for the follower. Also see section on *joining late*. **149**

What Invalidates the Salat

The salat is invalidated by:

1. Intentionally blowing air from the mouth. If it is unintentional, it counts as an addition, and one performs *sujud ba'di*.
2. Intentionally speaking invalidates the salat, unless it is **150**

126 This is the amount of time it normally takes to count the sunna dhikr on one's fingers and pray two or four sunna rak'as and come out of the mosque, i.e. 5-10 minutes.

necessary to correct the imam by words, which only invalidates the salat if excessive. If the speaking is unintentional, it counts as an addition, and one performs *sujud ba'di*. Clearing the throat, expectorating, sighing, burping, moaning from pain or crying from being overwhelmed with fear of Allah do not invalidate the salat.

151 3. Being prevented by something, such as nausea, or the need to pass wind, to vomit, urinate or the like, from performing an obligatory element of the salat, such as standing or bowing. But if one missed a sunna, it is recommended to repeat the salat within its time.
4. Breaking wudu' during the salat invalidates the salat, whether intentionally or unintentionally.
5. Unintentionally doubling the amount of rak'as (such as adding two rak'as to Subh, or adding four rak'as to Dhuhr, 'Asr, Maghrib or 'Isha).
6. Laughing out loud, intentionally or unintentionally, whether one is imam or a follower, but a follower who laughs must continue following the imam and then repeat the salat later.

152 7. Intentionally eating and/or drinking
8. Intentionally adding an extra prostration or a rak'a without reason
9. Intentionally swallowing vomit or regurgitated food
10. Realising one has five or less missed salats outstanding.[127] If one realises this in *jama'a*, one completes the salat, and then does what is outstanding, and finally re-does the salat just done in *jama'a*.

153 11. Realising, whilst praying, one has missed an obligatory element (such as a rak'a or sajda) in a previous obligatory salat.[128]
12. Realising, whilst praying, one missed the *qabli* prostrations of inattentiveness for missing three or more sunnas after one has left the mosque or a long time has passed, in a previous obligatory salat.[129]

127 If one realises that one has six or more salats outstanding whilst praying, his salat is valid.

128 This is because that previous salat is not completed and is outstanding, so the previous point number 10 applies in this case.

129 This is because that previous salat is not completed and is outstanding, so the previous point number 10 applies in this case.

Missing Obligatory Elements in Salat

If one realises during the salat that one has missed an obligatory element,[130] this element must be redone as described in the following scenarios, or the salat is invalidated: **154**

1. If one realises before standing up from the following ruku', do the missed element, regardless. For example, if whilst reciting the Fatiha in the second rak'a, one remembers a sajda was missed in the first rak'a, he should go straight into sajda, then stand and restart the Fatiha.
2. But if one realises after standing up from the following ruku', the entire previous rak'a is annulled. For example, if one has just completed the first rak'a, got up for the second, recited the Fatiha and a sura, gone into ruku' and, afterwards, remembers a sajda was left out in the first rak'a, the current rak'a is now counted as the first, because the first rak'a has been annulled. Two sajdas of inattentiveness should be done after the salam, for having added elements in the salat.
3. If one realises after the final salam that an obligatory element was **155** missed from a previous rak'a, that rak'a is annulled and one says *takbirat al-ihram* and completes the salat by adding a rak'a and performing the sajdas of inattentiveness after the salam for the added elements. However, if a long time passes, the salat is invalidated.

Doubt about Having Performed Obligatory Elements

If one is uncertain whether one has missed an obligatory element, one sticks to the minimum that one is sure of, does the assumed missed obligatory element and does two prostrations after the salam. **156**

1. If uncertain in Dhuhr or 'Asr whether one is performing the third or the fourth rak'a:
   - ☐ Always assume one is performing the third.
   - ☐ Add one rak'a (which will be assumed to be the fourth rak'a).
   - ☐ Do the sajdas of inattentiveness after the salam (to compensate

130 An obligatory element other than the *takbir al-ihram* and the intention, because without them the salat does not even start.

in case the rak'a was actually added in error).

2. If uncertain whether one performed one or two sajdas in the first rak'a after getting up for the second, one must assume only one sajda was performed and:
   - ☐ If one did not rise from ruku', one drops immediately into the assumed missing sajda and then gets up and restarts the second rak'a.
   - ☐ If one did rise from ruku', one treats the current rak'a as if it was the first (i.e. annulling the previous rak'a completely).
   - ☐ One prostrates after the salam in both cases to compensate for the added elements

This is only applicable to those who are not affected by obsession or constant doubts (*istinkah*). When somebody is obsessed and affected by constant doubts, and it happens to him at least once every day, he should ignore the doubts. In that case, if one doubts whether one has done three or four rak'a, one assumes four were made and only does two sajdas after the salam.

157 Sometimes it may happen that something is added and missed at the same time, so one must perform the sajdas before the salam. This is exemplified in this case:

1. If, after rising from ruku' of the third rak'a, one is uncertain whether one performed one or two sajdas in the second rak'a:
   - ☐ Treat the current rak'a as the second (i.e. annulling the previous rak'a completely, including the *sura* after the Fatiha).
   - ☐ Continue the current rak'a and complete the rak'a by sitting for *tashahhud*.
   - ☐ Add two more rak'as reciting in them the Fatiha only.
   - ☐ Do two sajdas of inattentiveness before the salam, for having added (i.e. the annulled rak'a) and subtracted something from the salat (i.e. the sura that was not recited in the third rak'a which became the second rak'a).

158 The sajdas are also performed before the salam if one realises in a obligatory salat, while rising, that one forgot to do the middle sitting. In this case, if, whilst rising, both hands and knees have left the ground:

- ☐ Continue to standing position, in order not to leave what is *fard* (standing) to perform a sunna (sitting).

- ☐ Perform the sajdas of inattentiveness before the salam (to compensate for the missing sitting, *tashahhud*, etc.).

Nevertheless, if both hands and knees have left the ground, but one returns to the sitting position,[131] one should do the sajdas after the salam, whether one did it out of ignorance of the above ruling, forgetfulness or intentionally.

If hands or knees have not left the ground yet, one sits back down and no sajdas are owed. But if one stands up regardless, one must do the sajdas before the salam for missing a stressed sunna element.

The above rules are only for the obligatory prayer. The *nafila* prayers, which are performed in units of two rak'as, have special rules:

1. If one realises while rising that one had forgotten the sitting for *tashahhud* one must sit back down, even if hands and knees left the ground, and then do the sajdas after the salam.
2. If – out of inattentiveness – one carries on, gets up and performs a third rak'a, one adds a fourth rak'a, and does the sajdas before the salam.

JUMU'A

Conditions to be Fulfilled for the Jumu'a Salat

The Jumu'a salat is *fard 'ayn* for whoever fulfils its conditions (*shurut ada'*). **159**
Its time begins at the time of Dhuhr and ends when there is just enough time for one rak'a of *salat al-'Asr* to be prayed before sunset after having completed Jumu'a.

The conditions to be fulfilled (*shart ada'*) in order for the Jumu'a salat to be valid are:

1. Permanent residence[132] (*istitan*) within a city or town or villages or settlements[133] if they can function independently and there are

131 Contrary to what he should do.

132 People who intend to stay in a place indefinitely.

133 This does not include caravans and tents and other types of settlements intended to be moved.

enough Muslim inhabitants to form a community

2. Two *khutbas* (speeches) immediately preceding the salat, separated by the imam sitting down, and each including:
   - □ Praising Allah by saying alhamdulillah
   - □ Prayers upon His Messenger ﷺ
   - □ Warnings of the Fire
   - □ Good tidings of the Garden and
   - □ At least one *ayah* of Qur'an

160 3. A *jami'* mosque in which to perform it, i.e. a properly constructed building, not a shack or an open field, etc.

4. An imam, who must be:
   - □ Free, i.e. not a slave
   - □ Male
   - □ Not travelling
5. A congregation (*jama'a*) of a minimum of thirty men who are permanent residents is necessary to initiate the Jumu'a. But once the Jumu'a is initiated, a minimum of twelve men who remain in the mosque until the end of the salat is required for the Jumu'a to be valid.

Conditions that Make Salat al-Jumu'a Obligatory

The conditions that make Salat al-Jumu'a obligatory (*shurut wujub*) are:[134]

1. Being a temporary resident (*muqim*)[135] if the Jumu'a has been originated by permanent residents
2. Not having a valid excuse, such as:
   - □ Being ill or tending to a sick or dying person Fearing for one's life, health or property if one goes and
   - □ Extreme weather conditions, such as heavy rain
   - □ Freedom, i.e. not being a slave
3. Living within three miles and a third (5.370 km) from the closest

134 These are conditions beyond a person's control.

135 A temporary resident or *muqim* is a person who intends to stay in a place for a period of time of 20 obligatory salats (i.e. four days) or more. This concept is different from the permanent resident (*mustawtin*), which is a person who intends to stay in a place permanently. The other status is the one of traveller (*musafir),* who is the person who intends to stay in a place for less than 20 obligatory salats.

mosque, if one lives outside of a city with a *jami'* mosque. If one lives within the city, to attend Jumu'a is obligatory even if one lives six miles (9.660 km) away from the *jami'* mosque.

4. Being male

Performing Jumu'a relieves one of the obligation of Dhuhr for those who are not obliged to attend, such as the traveller, the slave, the child, the woman, the one who lives out of range, the one who has a valid excuse - it is *mustahabb* for them to attend the Jumu'a prayer. 161

It is obligatory to attend the Jumu'a when the adhan sounds for the person close to the mosque, but the person further away must take into account the travel time.[136] This means that any action which distracts one from setting off for Jumu'a becomes haram, and business transactions within this time should be nullified.

Sunan and Mustahabb Elements of Salat al-Jumu'a

It is a sunna of the Jumu'a to perform a ghusl immediately before setting out for it, i.e. if one sleeps, eats lunch or the like after the ghusl and before setting out for the Jumu'a, ghusl should be repeated. 162

The *mustahabb* elements of the Jumu'a are:

1. To go at midday
2. To be well-dressed
3. To apply perfume (for men only)
4. To be well-groomed, i.e. to apply the elements of *fitra*:
    - To trim the moustache so the upper lip is free from hair
    - To keep the nails trimmed short
    - To shave off all pubic hair
    - To remove hair from the armpits
    - To brush the teeth

136 This means that the person far from the mosque must take into consideration the distance he has to cover in order to be able to attend the *Jumu'a*. Therefore to set off for the salat becomes for him obligatory even before the adhan takes place because of the extra distance he should cover in order to attend the salat.

JAMA'A

The Jama'a or Group Salat

163 A group of people (*jama'a*) is obligatory for *salat al-Jumu'a*, but for all other obligatory salats it is sunna to pray in a *jama'a*, which yields 25 to 27 times the reward[137] of praying alone, if one catches even a single rak'a with the imam (by doing ruku' with stillness before the imam rises from his ruku') unless one intentionally delayed joining the salat.

164 It is *mustahabb* to repeat an obligatory salat, previously done alone, behind an imam in a mosque (or elsewhere if the *jama'a* is a minimum of three people)[138] – except two salats:

1. *Salat al-Maghrib* and
2. *Salat al-'Isha* if one has prayed *Salat al-Witr*

The Imam

165 For the salat to be valid, the imam must fulfil the following conditions:

1. Be male. If a man or woman prays behind a woman[139] the salat is invalid.
2. Legally responsible (*mukallaf*). To follow a person who is insane, intoxicated or a child invalidates the salat. It is not permitted to make a child lead a *nafila* salat, but if it happens it remains valid.
3. Physically capable of performing the obligatory elements of the salat. The imamate of a physically incapable person over physically capable people is invalid. However, an incapable person can lead people with a similar incapacity, e.g. a person unable to stand can lead other

137 There is a discussion among *'ulama* whether praying in a mosque is a condition for the reward, this being the opinion of Ibn Hajar among others, but what is certain is that reward is added by walking to it and the increase in mercy is commensurate with the number of Muslims, etc.

138 Two people plus the one repeating the salat. It is not recommended to repeat the salat with one person only, unless that person is the permanent imam (*imam ratib*) of a mosque. Whoever repeats the salat with a *jama'a* should do it as a follower (*ma'mun*) not as an imam, otherwise his salat and the salat of those behind him will be invalidated.

139 There is difference of opinion (*ikhtilaf*) about women leading women, but the majority of 'ulama of all the madhhabs do not take this position.

people unable to stand.

4. Know enough fiqh and Qur'an to perform a valid salat, and also the rulings pertaining to wudu' and ghusl.

And the imam must not: **166**

5. Be a transgressor (*fasiq*), meaning that:
   - ☐ He performs wrong action, which has two levels:
     - If the wrong actions are not connected directly to the salat, the follower should repeat the salat in the *ikhtiyari* time, and one should not take him as imam.
     - If the wrong action invalidates the actual salat, it must be repeated.
   - ☐ He has a deviated 'aqida, which also has two levels:
     - If the scholars ('ulama) do not agree as to whether his beliefs constitute *kufr*, the follower should repeat the salat in the *ikhtiyari* time, and one should not take him as imam.
     - If his *kufr* is clear, all salats behind him are invalid and must be repeated.
6. Be a person who mispronounces the Qur'an (*lahhan*):
   - ☐ The person who unintentionally mispronounces – the salat behind him is valid, but it is haram to take him as imam if there is a person who can recite correctly.
   - ☐ The person who intentionally mispronounces – his salat and the salat behind him are invalid.
   - ☐ The person who is unable to pronounce and who cannot learn because of some defect – the salat behind him is valid, but it is haram to take him as imam if there is a person who can recite correctly.
   - ☐ The person who is able to pronounce correctly and does it – he should always be given precedence as imam.
7. Nor must the imam himself be a follower of another imam (i.e. a person who misses one or more rak'a with the imam cannot follow another latecomer after the actual imam has finished, as this invalidates the salat).
8. And specifically for the Jumu'a only:
   - ☐ He should not be a slave or
   - ☐ A traveler – because Jumu'a is not obligatory for these two.

167 It is *makruh* for the imam to be:

1. Incontinent (see section on *what breaks wudu'* in the Book on Purification) or leaking blood or pus from wounds or sores, if the follower does not suffer the same
2. A country dweller[140]
3. Disliked by the people in the *jama'a* because of a matter of his Deen, such as not paying his debts though he can afford to
4. 168 Paralysed in one of his limbs, limping (unless it is minor), or missing a limb
5. Without a cloak if leading the salat in a mosque

169 And it is *makruh* to:

1. Let the rows be broken up by columns and such, unless there is no space
2. Stand in front of the imam, unless there is no space anywhere else
3. And to make a *jama'a* after the regular imam (*ratib*) has already prayed

170 And the role of the imam must be given to a person who does not incur the disrespect or dislike of the *jama'a* he serves, as it is an elevated position which could generate jealousy and rumour.

So if he is *ratib*, i.e. appointed to lead the five daily salats, it is *makruh* that he is:

1. Unknown to the people he leads, i.e. whether he is upright (*'adil*) or a transgressor (*fasiq*)
2. Effeminate (*ma'bun*)[141]
3. Uncircumcised,[142] unless he has a valid excuse
4. A slave

---

140 According to the gloss of at-Talib ibn al-Hajj, the reason is that the country dweller does not normally perform *jama'a* or even *Jumu'a* in the city or village.

141 According to another interpretation *ma'bun* may mean homosexual. In this context, it will mean a person about whom there are rumours that some time in the past he engaged in homosexual practices, not that he is actually a homosexual since that would make him ineligible as imam.

142 According to the *mashhur* position of the *madhhab*, circumcision is a sunna. It is haram to show one's nakedness to another person unless it is necessary for medical reasons. Therefore, it is not halal to show one's nakedness in order to perform a sunna. Otherwise, circumcision is normally performed between the ages of seven and ten.

5. Castrated, meaning missing his penis, testicles or both
6. An illegitimate child born outside of wedlock

However, if there is no-one available free of the following defects, it is permissible for the imam to be: 171

1. Impotent (*'innin*)[143]
2. Blind
3. Incapable of pronouncing some letters
4. Leprous, if his affliction is slight

The Follower

The follower must follow the imam in everything related to the salat, unless the imam adds something to the salat, in which case two scenarios apply: 172

1. If the follower is certain the imam added something in error, he:
   - Must not follow (e.g. if a fifth rak'a is added to 'Isha, he stays sitting and waits for the imam to say the salam[144] or his salat is invalidated unless he follows due to inattentiveness or ignorance of this rule) and
   - Should say "*subhan'Allah*" to warn the imam of the error, and if he still fails to realise his mistake, the follower should verbally correct him.
2. If the follower knows, believes or assumes that the imam is doing that because he is making up for an invalid previous rak'a, he must follow the imam when he stands for a fifth rak'a. If he remains intentionally seated, his salat is invalidated, but if he remains seated due to inattentiveness, his salat remains valid.

If the imam is in doubt whether his addition is an error, he should correct it only if two reliable witnesses (*'udul*) or more warn him. But if he is certain he knows what he is doing, he ignores the two witnesses but not more witnesses than that.

Joining the Salat Late

143 Another meaning of the word *'innin* is the person who has such a small penis that he is incapable of performing sexual intercourse.
144 Then, he adds two prostrations after the salam.

173 1. The latecomer joins the imam immediately, uttering the obligatory *takbirat al-ihram* while standing, regardless of the imam's standing, bowing or sitting and

174 2. Adds another *takbir* only if the imam is in ruku' or sujud (missing this does not invalidate the salat) – but not if the imam is standing or sitting; in that case he only utters the *takbirat al-ihram.*

3. And immediately assumes the same position as the imam and then follows him to the end of the salat.

175 4. When the imam finishes his salat with salam the latecomer must rise and:
   - □ Make up whatever recitation he missed at its proper volume (*qada'*).
   - □ And carry on with the movements from where he joined (*bina'*).

So when joining late in *salat al-'Isha*, for example, where the latecomer catches the start of the last rak'a the following steps are taken:

1. The latecomer initiates his salat with *takbirat al-ihram*, and joins the imam's final rak'a in what is his own first rak'a, with silent recitation of the Fatiha, following the imam.
2. After the imam finishes with the salam, the latecomer stands up, does his second rak'a reciting Fatiha and a surah aloud, making up the recitation of the imam's first rak'a, which he missed.
3. Then he sits and says *tashahhud* at the end of this rak'a, because it is his second and as the previous sitting with the imam is not counted.
4. Then he stands up for his third rak'a, reciting the Fatiha and a sura aloud, making up the missed recitation of the imam's second rak'a but does not sit as it is his third rak'a.
5. In his fourth rak'a, he recites Fatiha silently without a sura because this is the recitation of the imam's third rak'a which he missed.
6. He then stays sitting, says the *tashahhud.* finishing with his own salam.

176 The latecomer's *takbir* after the imam's salam, depends on how many rak'as he performed with the imam:

1. If he got two rak'as (i.e. third and fourth rak'a of Dhuhr, 'Asr, and 'Isha or second and third of Maghrib), or he got less than one rak'a (i.e. he joined the last rak'a after the imam's ruku'), then he says

takbir when rising.

2. But if he got one rak'a (i.e. the third rak'a of Maghrib or the second rak'a of Subh), or he got three rak'as (second, third or fourth rak'a only of Dhuhr, 'Asr and 'Isha), then he does not say *takbir.*[145]

It is important to note that the latecomer to Subh does not do the *qunut* supplication.

Inadvertently leaving out any correctable mistakes by the latecomer (and follower) are covered by the imam as long he leads; but the latecomer is himself responsible for all errors made by him after the imam's salam.

However if the imam performs prostrations of inattentiveness, two scenarios 177
apply for the follower who has caught at least one rak'a with the imam:

1. If they are made before the imam's salam, the latecomer performs them with him, and then completes his salat as described in the example above, finishing with his own salam.
2. But if they are made after the imam's salam, the latecomer postpones them until after his own salam.[146]

The latecomer does these prostrations of inattentiveness even if he was 178
not present during the imam's mistake. But if he did not catch a rak'a with the imam, he should not do prostrations of inattentiveness with the imam (whether *sujud qabli* or *ba'di*). If he does, his prayer will be invalid.

Appointing a Replacement Imam

The invalidation of the imam's salat always invalidates his followers' except 179
in two cases:

145 This is because the follower has already uttered a *takbir* in order to stand up, but he had to remain seated because he has to follow the imam. This is the *mashhur* (widespread) position. The opinion of Ibn Majishun is to do the *takbir* in all cases. In this case, it is better for ordinary people to use the opinion of Ibn Majishun so that they do not become confused in this matter.

146 If one prostrated *ba'di* with the imam out of ignorance of the ruling or intentionally, one's salat is invalid. If one did it by mistake, the *salat* is valid, but one repeats the prostrations of inattentiveness after one's own salam.

180 1. If the imam suddenly realises that he is not in a state of ritual purity or
2. He is overcome and breaks his ritual purity

– on condition that he immediately leaves the salat.[147]

181 And then it is *mandub* that he appoints a new imam from among the followers to finish the salat; and if he does not, the followers may either:

1. Finish their salats individually or
2. Send one of them forward as the replacement imam who will finish the salat.[148]

147 If he does not leave the salat immediately, his salat and the ones following him will be invalidated, because they will then be following an imam who is in ritual impurity (*muhdith*).
148 And this is the only choice in *Jumu'a*, as congregation is a condition for the validity of *Jumu'a*.

Zakat

Zakat

INTRODUCTION TO ZAKAT

THE ROOT MEANING of the word 'zakat' is 'increase' because it increases the blessing of the wealth paid, making it an increase.

The Shari'ah meaning of zakat is 'a specific amount of wealth taken from a specific kind of wealth when it reaches a specific amount, in a specific time, which is then given to specific groups of people.'

Zakat is fard according to the Book of Allah and the Sunna of His Messenger ﷺ and denying this is kufr.

A Muslim who refuses to pay it while confirming it is fard, is fought until he gives it and then punished for withholding it.[149]

There are two kinds:

1. *Zakat al-amwal* on wealth, which is paid with the passing of a lunar year
2. *Zakat al-abdan* or *zakat al-fitr*, which is connected to the 'Id at the end of Ramadan

CONDITIONS OF OBLIGATION

The seven conditions of obligation (*shart wujub*), which are not asked for because they are out of the *mukallaf's* control, are:

1. Being Muslim
2. Free (not a slave)
3. That the wealth has reached the *nisab* (minimum amount upon

149 This means that the existence of Islamic governance is necessary for proper collection and distribution of zakat.

which zakat is levied)

4. That the wealth is in one's legal possession (i.e. not stolen or usurped)
5. That the wealth has been in one's possession for the duration of one lunar year (except for crops)
6. As for the zakat on livestock, the arrival of the collector
7. As for the zakat of money, to be free of debt

CONDITIONS OF VALIDITY

There are five conditions for the validity of zakat:

1. Having the intention to pay it
2. Paying it after it becomes obligatory (i.e. it cannot be paid in advance)
3. Paying it to an amir, on the condition that he is just, i.e. he distributes it to the right recipients[150]
4. To give it directly to the eight recipients if there is no amir
5. It should be paid in kind, meaning zakat on crops is paid in crops, zakat on gold is paid in gold, etc. except for shops and their like, whose zakat is paid for in their equivalent in gold or silver

WEALTH SUBJECT TO ZAKAT

182 It is obligatory to give zakat from three kinds of wealth:

1. Money (*Zakat al-'ayn*) which includes:
   - □ Gold and silver[151]
   - □ Mined products and
   - □ Trade merchandise
2. Crop harvests (*Zakat al-harth*) which include:
   - □ Grain
   - □ Certain fruits
3. Livestock (*Zakat al-mashiya*) which include:
   - □ Camelids (e.g. camels)
   - □ Bovines (e.g. oxen, cows, etc.)

---

150 If an unjust amir takes the zakat by force, the obligation is lifted from the individual, even if the amir uses the funds for his own purposes. But one should avoid giving it to him if possible.

151 Al-Marrakushi says in his commentary of the *Murshid al-Mu'in* that zakat should be paid on what has supplanted gold and silver nowadays, i.e. paper money, if its value reaches the *nisab*.

- □ Ovines and caprines (i.e. sheep and goats)

TIME OF ZAKAT

As for the time the zakat is due, there are two rulings depending on the kind of wealth: 183

1. As for money and livestock, it is a condition that one full lunar year passes.
2. Crops have two rulings, depending on the type:
   - □ Grain is due when it is ready to be harvested, i.e. when it may be separated from the ears. Grains include:
     - Wheat
     - Barley
     - *Sult* (a grain in between wheat and barley)
     - *'Alas* (a kind of wheat)
     - Rice
     - Millet and
     - Corn
   - □ The rest of the zakatable agricultural products are due when they ripen and divide into four categories: 184
     - Dates
     - Raisins
     - Oil-producing plants (which are paid in oil) which include:
       - ⊙ Olives
       - ⊙ Sesame seeds
       - ⊙ Red radish seeds and
       - ⊙ Safflower
     - And seven pulses, which include:
       - ⊙ Lentils
       - ⊙ Broad beans
       - ⊙ Chickpeas
       - ⊙ Normal beans
       - ⊙ Lupins
       - ⊙ Peas and
       - ⊙ Grass peas (*julubban*)

Except oil-producing plants, all crops are paid in kind. There is no zakat on green vegetables and fruit (except dates, raisins and sultanas).

MINIMUM AMOUNT (NISAB) AND PAID AMOUNT

185 On all the above-mentioned crops two rulings apply:

1. 10% Is paid if the irrigation is natural.
2. But only 5% is paid if artificial irrigation was used.

186 The *nisab* is the minimum amount one must possess in order for zakat to be obligatory. On these crop products the *nisab* is five *wasqs*.[152] The *nisab* on grains is calculated after it is dried and sifted. The *nisab* on fruits is calculated after they are dried.

Money can be in the form of:

1. Silver, for which the *nisab* is 200 Islamic dirhams, each being the size of 50 and ²⁄₅ of a grain of medium sized barley (i.e. approximately 600 g of pure silver)
187 2. Gold, for which the *nisab* is 20 Islamic dinars each being the size of 72 grains of medium sized barley (i.e. approximately 88 g of gold)

A quarter of a tenth i.e. 2.5% is paid on both gold and silver. It is permissible to pay the zakat on silver in gold and vice versa, according to the exchange rate of the time.

Trading merchandise divides into two categories:

188 1. Continuous trade (*idara*), in which merchandise is constantly bought and sold, as with shopkeepers, wholesalers, etc., who must take into account the following when calculating the *nisab*:
    - □ The value of the trade merchandise is calculated from the total sale price of stock at the end of the fiscal year. Its value should be calculated according to the equivalent amount of gold or silver.

152 One *wasq* equals sixty *sa'*s. A *sa'* equals four *mudds* and a *mudd* is the double cupped handful of the Prophet ﷺ equivalent to approximately two litres. Therefore, the *nisab* of crops is 300 *sa'*s, or approximately 600 litres.

- □ This is added to any other monetary wealth and if it reaches the *nisab*, 2.5 % is paid of the combined value in gold or silver.
- □ In addition the trader adds the value of commercial debt owed to him, if the debt is due and he expects it to be paid.

2. Speculative trade (*ihtikar*) is trade of merchandise bought to be sold when the market price rises, such as real estate. One only pays zakat when the property is sold and the money received, on the condition one year passed since the property was put up for sale.[153] 189

Speculative traders pay no zakat on debts owed until they are paid; then zakat is due immediately if it was owed for more than a year, and amounts to the *nisab* or more.[154 155]

As for goods that are not objects of trade, such as the house in which one lives, one's car, furniture, utensils, etc., they are not subject to zakat.

ZAKAT ON LIVESTOCK

Zakat is levied on three types of livestock,

1. Camelids
2. Bovines and
3. Ovines – and there is no difference whether or not these animals are used for work, or whether they are fed or left to graze

Camelids

Camels (including dromedaries, lamas, etc.), for which the *nisab* is five head,[156] are counted as follows: 190

153 e.g. if a house has been sold six months after being put up for sale, one waits an additional six months; if any money above the *nisab* is left at that time, zakat is paid on it. If less than the *nisab* remains, it is added to whatever other money was possessed for a full year.

154 Or it amounts to the *nisab* or more when combined with money already possessed or to be possessed within that fiscal year.

155 As for the speculative trader, the payment of both the sale of a property or the settlement of a debt should be in money (*'ayn*).

156 If less than five, one pays nothing.

	Number of camels owned	Payment
	5-9	1 one-year old goat or sheep (*jadha'a*)
	10-14	2 one-year old goats or sheep (*jadha'a*)
	15-19	3 one-year old goats or sheep (*jadha'a*)
	20-24	4 one-year old goats or sheep (*jadha'a*)
191	25-35	1 one-year old she-camel (*bint makhad*)
	36-45	1 two-year old she-camel (*bint labun*)
192	46-60	1 three-year old she-camel (*hiqqa*)
	61-75	1 four-year old she-camel (*jadha'a*)
193	76-90	2 two-year old she-camels (*bint labun*)
	91-120	2 three-year old she-camels (*hiqqa*)
194	121-129	3 two-year old she-camels (*bint labun*) or 2 three year-old she-camels (*hiqqa*)
195-6	130 and above	For every extra 50 camels, 1 three-year old she-camel (*hiqqa*) and for every extra 40 camels, 1 two-year old she-camel (*bint labun*)[157] [158]

157 The following examples are taken from *al-Fiqh al-Maliki wa Adillatuhu* by al-Habib bin Taher: 130 head is 1 *hiqqa* and 2 *bint labun* (i.e. 50+40+40=130), 140 head is 2 *hiqqa* and 1 *bint labun* (i.e. 50+50+40=140), 150 head is 3 *hiqqa* (50+50+50=150), 160 head is 4 *bint labun* (i.e. 40+40+40+40=160), 170 head is 1 *hiqqa* and 3 *bint labun* (i.e. 50+40+40+40=170), 180 head is 2 *hiqqa* and 2 *bint labun* (i.e. 50+50+40+40=180), 190 head is 3 *hiqqa* and 1 *bint labun* (i.e. 50+50+50+40=190), 200 head is either 4 *hiqqa* or 5 *bint labun* (i.e. 4x50=200 or 5x40=200).

158 Whenever it is mentioned that the animal is one-year old, it means that the animal has completed a year of life and has entered its second year of life. A two-year old animal, means that it has completed two years of life and has entered in its third year of life. And so on.

Bovines

As for bovines (including cows, oxen, etc.), nothing is paid for less than thirty head. Their zakat is calculated in the following manner: 197

1. For every 30 head, one must give away one two-year old calf (*tabi'*).
2. For every 40 head, one must give away one three-year old female calf (*musinna*).

And so forth. 198

See an example of this below:

Number of bovines owned	Payment
30 – 39	1 two-year old calf (*tabi'*)
40 – 59	1 three-year old female calf (*musinna*)
60 – 69	2 two-year old female calf (*tabi'*)
70 – 79	1 two-year old female calf (*tabi'*) + 1 three year-old female calf (*musinna*)
80 – 89	2 three-year old female calves (*musinna*)
90 – 99	3 two-year old calves (*tabi'*)
100 – 109	2 two-year old calves (*tabi'*) + 1 three-year female calf (*musinna*)
110 – 119	1 two-year old calf (*tabi'*) + 2 three-year old calves (*musinna*)
120 – 129	4 two-year old calves (*tabi'*) or 3 three-year old female calves (*musinna*)

Sheep and Goats

As for sheep and goats, for which the *nisab* is forty head, they are counted as follows:

	Number of sheep and goats owned	Payment
	40 –120	1 one-year old goat or sheep (*jadh'a*)[155]
199	121 – 200	2 one-year old goats or sheep (*jadh'a*)
	201 – 399	3 one-year old goats or sheep (*jadh'a*)
200	400	4 one-year old goats or sheep (*jadh'a*)

From 400 onwards add a one-year old goat or sheep (*jadh'a*) for every 100 head. So zakat for 500 head will be 5 goats or sheep, for 600 head zakat will be 6 goats or sheep, and so on.

Studs and animals that are fattened, ill or handicapped are not taken for zakat; nor are pregnant females, young, old or rearing females. In general, neither the best nor the worst are taken, but rather what is between.

THE ZAKAT FISCAL YEAR

201 The rulings related to the zakat fiscal year depend on the type of wealth:

1. The beginning of the fiscal year for profits is identical to the beginning of the fiscal year for the capital whether the original amount is above the *nisab* or not, as illustrated below:

    - □ Whoever owns 20 Dinars for 10 months, then uses them to buy merchandise and after 2 months sells it with a profit of 30 Dinars, must pay zakat on the 30 Dinars immediately, as if the

159 Male or female.

profit had been hidden in the original amount of 20 Dinars.

- Whoever owns 15 Dinars for 10 months, then uses them to buy merchandise and after 2 months sells it with a profit of 30 Dinars, must pay zakat on the 30 Dinars immediately, as if the profit had been hidden in the original amount of 15 Dinars.

2. The beginning of the zakat fiscal year for livestock is the same as that for the animals' parents, but has special rules related to the type of increase mentioned in point B. below:

- A. If increase was caused by birth, one must pay at the end of the fiscal year, as if the full size of the flock had been hidden in the original amount, as illustrated in the following examples:

  - Whoever owns 80 sheep (for which the zakat is one sheep) which subsequently give birth to 41 lambs, bringing the flock to 121 head by the end of the fiscal year, must pay the zakat (two sheep) for the new total amount.

  - Whoever owns 30 sheep (which is below the *nisab*) which subsequently give birth to 10 lambs, bringing the flock to 40 head by the end of the fiscal year, must pay the zakat (one sheep) for the new total amount, as if the full size of the flock had been hidden in the original amount of 30 sheep.

- B. However for any increase in livestock not caused by birth (such as acquisition through inheritance, gifts or purchase), special rules apply, as illustrated below:

  - If the original number of head is below the *nisab* and the *nisab* is reached by acquisition of additional head, one does not pay zakat until a full year has passed. e.g. whoever owns 20 sheep (which is below the *nisab*) and acquires an additional 20 head, does not pay zakat until a full year has passed on the total of 40 head.

  - But if the original number of head is above the *nisab* and the next level of zakat is reached by acquisition of additional head, one must pay the full zakat immediately, e.g. whoever owns 40 sheep (for which the zakat is one sheep) and acquires an additional 81 head, he must immediately pay the full zakat for 121 head (for which the zakat is two sheep), because his original flock was above the *nisab*.

EXCEPTIONS

202 As for an intermediate amount (*waqs*), no payment is made, as seen in the following three examples (also see tables in section on *Zakat on livestock*):

1. Whoever has seven camels only pays zakat for the first five head and the remaining two are exempt, as they do not reach the next level, which is ten head.
2. Whoever has between 40 to 59 cows only pays one calf.
3. Whoever has between 40 to 120 sheep or goats only pays one sheep or goat.

This ruling only applies to livestock. As for money and goods, zakat is paid on everything above the *nisab*, regardless of how small the amount.

There is no zakat on any amount below the *nisab* in any of the three kinds of wealth.

203 Honey, fruit (other than dates and grapes) and vegetables (other than certain string beans) are exempt from zakat, since zakat is only due on grain and fruits on two conditions:

1. That they are a basic source of nourishment and
2. That they can be stored, e.g. rice, wheat and corn

GROUPED CATEGORIES

204 It is not a condition that the type of wealth be the same to reach the *nisab*, rather the minimum threshold for cash currency can be achieved from both types:

1. As for money, the *nisab* can consist of:

- ☐ Only gold, such as 20 dinars or
- ☐ Only silver, such as 200 dirhams, or
- ☐ A combination of gold and silver, such as 10 dinars + 100 dirhams, or 5 dinars + 150 dirhams or 15 dinars + 50 dirhams, etc.

2. As for livestock, the *nisab* can consist of: **205**
   - ☐ 40 Sheep or
   - ☐ 40 Goats or
   - ☐ A combination of goats and sheep, such as 20 goats and 20 sheep. Similarly, dromedaries may be mixed with Bactrian camels and cows may be mixed with buffalo, etc.
3. As for harvested crops, the *nisab* can be reached by: **206**
   - ☐ A combination of three types of grain:
     - Wheat
     - Barley
     - *Sult* (a grain between wheat and barley)
     - A combination of the seven pulses (such as peas, broad beans, etc.)
   - ☐ A combination of raisins of different types (such as sultanas, red, black, etc.)
   - ☐ A combination of dates of different types

RECIPIENTS OF ZAKAT

The legitimate recipients constitute eight groups, as mentioned in Surat at-Tawba (*ayat* 60):

1. The poor (*fuqara*), meaning those who do not have food for themselves and their dependants for one year[160] **207**
2. The destitute (*masakin*), meaning those without the means to support themselves even for a single day

   Whoever claims to be poor, this claim is accepted, unless his appearance contradicts it. If he claims to have dependents, it demands verification. If he is known to be rich, he must prove he has lost his money.

   And these two groups must meet the following four conditions; they

160 *Aqrab al-Masalik* by ad-Dardir

must be:

- □ Free, i.e. not slaves
- □ Muslims
- □ Not descendants of the Messenger of Allah, ﷺ, but if such descendants are not supported by the Muslim common fund (*bayt al-mal*), as is the case today, they may receive zakat[161]
- □ Not already being maintained by a responsible relative, such as:
    - A wife being maintained by her husband or
    - A son by his father, or
    - A poor father by his son, or
    - If a non-responsible person has taken on the responsibility, such as a stepdaughter being maintained by her stepfather

3. Fighters in the way of Allah (*fi sabilillah*) for:
    - □ Equipment
    - □ Weapons and
    - □ Transport to the front lines
4. To use the zakat funds to buy and set free Muslim slaves (*riqab*)
5. The collectors and distributors of zakat (*al-'amilin 'alayha*), who must be:[162]
    - □ Muslim
    - □ Free (i.e. not slaves)
    - □ Not family of the Messenger of Allah ﷺ
    - □ Upright (*'adl*) and
    - □ Know the fiqh of zakat[163]
6. People in debt (*gharimin*), i.e. if one has contracted a halal[164] debt with another person, one will be given from the zakat funds towards the complete payment, after one has paid whatever one could.

208 7. Those whose hearts are to be reconciled (*al-mu'allafatu qulubuhum*), i.e. unbelievers (*kuffar*) encouraged to enter Islam by receiving funds, according to the strongest opinion. There is another opinion that this refers to new Muslims, who are given zakat funds in order to

161 *Ash-Sharh as-Saghir* by ad-Dardir

162 Shaykh Mayyara includes being adult and male in his commentary, but those conditions are not mentioned in other reliable books of fiqh like the *Mukhtasar*, *Aqrab al-Masalik*, etc.

163 The *Mukhtasar* of Khalil mentions points d. and e.

164 I.e. not contracted by gambling, selling alcohol or any other type of haram transaction.

strengthen their Islam.
8. Travellers (*ibn as-sabil*) who do not have the means to return to their homeland, provided their trip was for a halal purpose

Zakat is never used to purchase or build mosques, vehicles, freeing prisoners, publishing books, etc.

ZAKAT AL-FITR

Zakat al-Fitr is obligatory for all able Muslims and consists of one *sa'*[165] per person, for themselves and each of their Muslim dependents (e.g. children, wives, poor parents, etc.). **209**

It is given in the staple food normal to the people of the area from: **210**

1. Wheat
2. Barley
3. *Sult*
4. Dates
5. Dried cheese
6. Raisins
7. Millet
8. Corn
9. Rice[166]

It is recommended to give it after Fajr before going to the place of the 'Id salat. It is permitted to give it one or two days before the 'Id. It is haram to delay it until *Maghrib* of the day of the 'Id, unless one has not found a poor person to hand it to. If not paid on time, *Zakat al-Fitr* remains a debt.

It is given to the destitute (*miskin*), or poor (*faqir*) free Muslim (see above). More than one *sa'* may be given to one *faqir* or it may be split among several.

The wisdom is that everybody on the day of 'Id is free from the need to beg for food.

---

165 A *sa'* equals four *mudd* which is the double cupped handful of the Prophet ﷺ equivalent approximately to two litres.
166 Note that Abu Hanifa's School allows paying the equivalent value of *Zakat al-Fitr* in money.

Fasting

Fasting

FASTING AND ITS LEGAL CATEGORIES

THE LINGUISTIC MEANING of Fasting – '*sawm*' or '*siyam*' in Arabic – relates **211-2**
to 'abstaining,' but within the context of the Shari'a it means to abstain from food, drink and sex between dawn (Fajr) and sunset (Maghrib) with the intention of drawing nearer to Allah.

1. There are five fasts which are obligatory in total:

    - ☐ All the month of Ramadan
    - ☐ Making up any days of Ramadan and other obligatory fasts if missed for any reason
    - ☐ Fasting the expiation (*kaffara*) for transgressions that demand it, such as intentionally not fasting during Ramadan without a valid excuse, or unintentionally killing a human being
    - ☐ Vowing by Allah to perform a fast, such as saying: "By Allah, tomorrow I will fast." It is obligatory to fast that day (and this is true for all forms of worship)
    - ☐ Pronouncing *dhihar*[167] against one's wife; one must fast two consecutive lunar months before one may have intercourse with her

2. Ibn 'Ashir mentions four recommended fasts:

    - ☐ In the month of Rajab
    - ☐ In the month of Sha'ban
    - ☐ The whole month of Muharram, but especially the 10th, the day of 'Ashura
    - ☐ The first nine days of Dhul Hijja especially the 8th and 9th,

---

167 *Dhihar* is to equate one's wife with a *mahram* family member, such as a mother or sister, alleging by that she has become sexually impermissible to the husband.

the day of 'Arafa, except for pilgrims on hajj (*hajjis*)

3. Voluntary (*nafila*) fasts may be on any day, except those on which fasting is forbidden or obligatory

Fasting the month of Ramadan is necessarily known to be one of the five pillars of Islam and anyone denying this is a disbeliever (*kafir*). A Muslim refusing to do it, while confirming it is obligatory, is liable to a corrective punishment by the authority.

DETERMINING THE TIME OF RAMADAN

213 The time of Ramadan is determined in one of two ways:

1. If sighting the new moon on the 29th of Sha'ban is confirmed, the following day is counted as the 1st of Ramadan.
2. If not, one more day (i.e. the 30th of Sha'ban) is added, and the following day will be counted as the 1st of Ramadan.

Both demand knowledge of the Islamic calendar, as the start of Sha'ban must be known. Ramadan is not established by astronomical calculations, but by moon-sighting.

There are two ways to confirm that the new moon has been seen:

1. Two trustworthy witnesses testify to the relevant authorities.
2. If many people (i.e. too many to make it inconceivable that they have conspired to lie) testify together to the relevant authorities.

The relevant authority is the Khalifa, Sultan, Amir or a *qadi* appointed by one of them. The authorities may then spread the news in any way, but must confirm the sighting as above.

Trustworthy witnesses must be *mukallaf*, free, male, trustworthy, know the fiqh of the matter they are dealing with and in this case appointed by the authorities to avoid false reports.

If a single person sees the new moon, even if no-one else has, it is obligatory for him alone to fast.

OBLIGATIONS AND CONDITIONS OF FASTING

There are five obligations of fasting: **214-6**

1. To make one's intention after sunset (Maghrib) and before dawn (Fajr) to fast. If the intention was made after Fajr, that day's fast is invalid.
2. To avoid all sexual acts. The fast is not broken by ejaculation/orgasm because of a wet dream during daytime sleep, nor unintentionally issuing pre-ejaculatory discharge while awake. Nevertheless, the intentional issue of pre-ejaculatory discharge (*madhy*) during the day breaks the fast.
3. To avoid eating, drinking and smoking. Swallowing one's own spit and phlegm does not break the fast. Injections are disliked (*makruh*), but do not invalidate the fast.
4. To avoid intentional vomiting. Also, if some unintentional vomit is deliberately swallowed, that breaks the fast.
5. To avoid anything entering the stomach or passing the throat through the nose, eyes or ears via the Eustachian tube. Therefore, swimming, sniffing water deeply for wudu', putting drops in the eyes, using *kohl* and such should be avoided.

All fasts are from Fajr to Maghrib.

Fasting Ramadan is obligatory on six conditions, these being: that one is

1. Muslim
2. Sane
3. Adult
4. Healthy
5. Resident, i.e. not travelling
6. Being free from menstruation and post-natal blood

IMPEDIMENTS TO FASTING

There are two impediments to fasting: **217**

1. Being insane or having passed out (which does not include normal sleep):
   - □ When Fajr arrives, in which case the fast is not valid, and it must be made up if the fast is obligatory.
   - □ For a period between Fajr and Maghrib that lasts more than half the day, in which case the day must be made up if the fast is obligatory. But if the condition lasts half the day or less the fast remains valid.
2. Menstruation or post-natal bleeding any time between Fajr and Maghrib. There is no fast that day and it must be made up, if it is an obligatory fast. This prevention is lifted when bleeding is reduced to a white or transparent liquid with no tinge of pink, brown or yellow.

DISLIKED ELEMENTS OF FASTING

218-20 The disliked (*makruh*) elements of fasting are:

1. Fantasising, looking, kissing, touching or petting with sexual intentions,[168] even if one is sure one will not issue pre-ejaculatory discharge. It is forbidden if one is unsure.
2. Tasting food. Toothpaste should also be avoided, because it may be swallowed.
3. Lots of unnecessary talking.

The exceptions which do not break the fast are:

1. Unintentional vomiting
2. Accidentally swallowing insects which enter the mouth
3. And it is the same with dust from the road or from one's work, such as carpentry
4. Using a *siwak* (a teeth cleaning twig) on the condition that it is dry
5. And waking up in a state of *janaba* (major ritual impurity) after Fajr

168 These are *makruh* in Ramadan within normal permissible relations such as one's wife, but with other women they are haram both within and outside of Ramadan.

RENEWAL OF INTENTION

When fasting consecutive days, making one's intention before Fajr of the first day for all the days is enough, if the succession of fasting days is obligatory, as in Ramadan. But if the continuity of the fasts is interrupted for any reason (such as due to sickness or travelling), the intention must be renewed. **221**

And it must be renewed every night when fasting supererogatory fasts, even in succession.

RECOMMENDED ELEMENTS OF FASTING

Ibn 'Ashir mentions two sunna of fasting: **222**

1. To break the fast quickly after Maghrib with water and dates. If one eats without being sure it is Maghrib and:
   - □ Never finds out for sure, redo that day with no further expiation
   - □ Finds out it was, do nothing
   - □ Finds out it was not, redo that day with no further expiation
2. To postpone the *suhur*[169] meal (it may be as little as a glass of water) till shortly before Fajr. The Messenger ﷺ used to stop eating before Fajr for the time that it takes to recite approximately fifty *ayats*. If one eats because one is unsure whether it is Fajr or not and:
   - □ Never finds out for sure, redo that day with no further expiation
   - □ Finds out it was not, do nothing
   - □ Finds out it was, redo that day with no further expiation

Upon hearing the adhan for Fajr, stop eating.

BREAKING AN OBLIGATORY FAST

If breaking any obligatory fast (intentionally or not), that day must be made **223-4**

169 Meal eaten before Fajr (dawn).

up.[170] Intentionally breaking an obligatory fast, without a valid excuse (such as illness, etc.) is forbidden and demands *tawba.*[171]

If the fast is intentionally or unintentionally broken during Ramadan, it should be continued (unless there is a valid reason not to fast) due to the sacredness of the month, and that day must be made up as well.

Consecutive obligatory fasts (like the expiation of *dhihar*[172] or of Ramadan) should be continued if broken unintentionally, and that day must be made up as well.

If broken intentionally, then they should not be continued, because consecutiveness has been broken and the whole expiation needs to be started from the beginning.

Obligatory non-consecutive fasts (like the making up of the days of Ramadan, the expiation of an oath, etc.) do not need to be continued if broken, because it is obligatory to make them up anyway.

Five actions during Ramadan require expiation if fasting is intentionally broken and without any valid excuse,[173] [174] and these five actions during Ramadan[175] are:

1. Eating
2. Drinking
3. Smoking
4. Emitting semen (even if it was achieved only by fantasising)
5. Abandoning the intention to fast altogether

225 Except in three cases:

1. Making a wrong interpretation, on the condition that it was:
    - □ Plausible, such as:
        - The person who thinks that he can continue eating

170 Except if the person vows to fast a specific day and he breaks the fast due to illness, menstruation or passing out, then he does not have to make up that day, because the specific date has been missed. He will still have to make up the day if he broke his fast due to forgetfulness, travelling or just willingly and without any valid reason.

171 Turning to Allah in repentance.

172 See note 167.

173 This means that intentionally breaking the fast in Ramadan due to a valid excuse such as illness, menstruation, etc. does not require expiation.

174 This means that unintentionally breaking the fast in Ramadan does not require expiation.

175 This means that breaking the fast in a month other than Ramadan does not require any expiation.

after he broke the fast out of forgetfulness
- Breaking the fast when travelling less than minimum distance (see below)
- A woman whose menstruation ended before Fajr, but did not do ghusl until after Fajr believing she should not fast

☐ If the interpretation is implausible, then this demands expiation. Examples of implausible interpretations include:
- Reacting to the hadith "There is no fast for the one who backbites," by believing one may eat because one has been backbiting
- Not fasting because one expects a habitual medical condition later in the day that allows fast breaking

2. Health issues including:
☐ Old age.[176] In this case it is recommended to give a poor person a *mudd*[177] of grain for every missed day, but the missed days are not made up.
☐ Pregnancy. In this case the missed days must be made up.
☐ Breastfeeding.[178] In this case it is obligatory to give a poor person a *mudd* of grain for every missed day and the missed days must be made up.
☐ Suffering hardship because of weak health (even if able to endure fasting)
☐ Fearing one will contract an illness
☐ Worsening an illness
☐ Slowing the recovery from an illness. In this case the missed days must be made up.[179]

Fearing death or extreme harm, such as loss of limbs, capacities or senses, makes fasting forbidden.

176 This applies when fasting entails for the old person a greater strain that it entails for the rest of the people.

177 A measure consisting in the amount of grain that can be held by both hands cupped together.

178 Pregnant and breastfeeding women are allowed not to fast when they fear for their own or their children's health, not just for being pregnant or breastfeeding.

179 All dispensations of fasting due to health reasons should be based on consulting a trustworthy doctor or on one's own experience.

3. Travelling that allows one to shorten the salats[180] (see *sunan of salat*) though it is recommended to fast. When a traveller arrives at his destination during the day, he need not start fasting, but should not eat in public during Ramadan. A person is not considered to be travelling until leaving the limits of his town. All days not fasted due to travelling must be made up.
   - □ In order to make permissible the breaking of the fast whilst travelling, the traveller has to set out before dawn with the intention of not fasting. If the traveller makes an intention to fast and sets out before dawn, it is not permitted to break it, unless there is a compelling need. If it is broken, the traveller will have to make up that day and expiate.[181]
   - □ The traveller will also be obliged to make up the day and expiate if he makes intention not to fast and sets out after dawn.[182]
   - □ If the traveller makes an intention to fast and sets out after dawn, he is not allowed to break his fast. Nevertheless, if he does he should only make up that day.

BREAKING A RECOMMENDED FAST

226 As for breaking a recommended fast:

1. It is forbidden to break it intentionally without a compelling need. If that happens, the fast does not need to be continued, but it must be made up.
2. It is allowed to break the fast if one's parents or teacher order us to break it out of compassion. In this case, it will be obligatory to make up the day.
3. Unintentionally breaking it does not demand making up the day, but the fast must be continued.
4. If it is intentionally broken out of a compelling need, the day does not need to be made up.

---

180 A permitted trip of 48 miles or 78 km one way.

181 Because having had the choice of intending to fast or not to fast, he deliberately chose to fast. Therefore, if he breaks the fast without a valid reason, he deserves the punishment of expiation. Ashhab, the student of Malik, said that expiation is not necessary, only the making up of the day.

182 Because dawn arrived and he had abandoned his intention to fast.

EXPIATION – KAFFARA

The expiation mentioned above can be done by one of these three actions: **227-8**

1. Fasting two lunar months consecutively[183]
2. Freeing a Muslim slave who is free of defects
3. Giving sixty free, poor Muslims a *mudd*[184] of staple food each, which is preferred by the Maliki 'ulama

183 If the consecutive fasting of the expiation is broken unintentionally, it must be continued immediately. If it is broken intentionally without a valid reason (i.e. illness, menstruation, coercion), the fast must be started over again.

184 Ashhab said that it is valid to replace one *mudd* of food with a lunch and a dinner.

Hajj

Hajj

INTRODUCTION TO HAJJ

HAJJ (PILGRIMAGE) IS one of the five pillars of Islam, and denying this is *kufr*. The word means 'intending to reach a destination.' Its time is from the 1st of the month of Shawwal, all of the month of Dhul-Qi'da and it ends on the 10th of Dhu'l-Hijja.

CONDITIONS OF HAJJ

The hajj is obligatory once in the lifetime on every Muslim who is: **229**

1. *Mukallaf* (hajj by a child is a *nafila*, and does not relieve him of the obligation)
2. Free (not a slave)
3. Able to do it, meaning:
   - □ Having the health, i.e. the physical strength to reach Makka
   - □ The wealth, i.e. money to cover one's travel expenses
   - □ Access to Makka, i.e.
     - One's physical and/or
     - Financial integrity is not threatened
     - And finally that it is possible to perform the obligatory salat with all its *fard* elements while travelling.

Performing an additional hajj after the obligatory hajj is *mandub*.

THE ELEMENTS OF HAJJ

The rituals of hajj are divided into three categories:

1. Pillars (*arkan*), non-performance of which cannot be mended by sacrificial slaughtering (*hady*)
2. Obligations, non-performance of which can be mended by sacrificial slaughtering

3. Sunna and *mustahabb* elements, non-performance of which demands no expiation

THE PILLARS OF HAJJ

There are four pillars (*arkan*) of hajj:

230 1. *Ihram*, which is the state of sacredness caused by the intention to go on hajj. The *ihram* garments themselves are not a pillar, but an obligatory element.
2. *Sa'y*, which is going back and forth between the hills of Safa and Marwa
3. *Wuquf fi 'Arafa*, which is to be present on the plain of 'Arafa during part of the night of the 10th of Dhu'l-Hijja, which in Shari'a terms is the time between Maghrib and Fajr that precedes the day
4. *Tawaf al-Ifada*, which is to go around the Ka'ba, after having visited Mina

SACRIFICIAL SLAUGHTERING

231 Non-performance of obligatory elements of the hajj which are not pillars can be mended by sacrificial slaughtering (*hady*). The meat is distributed to the poor, and comprises three types of animal in order of preference:

1. A camel
2. A cow
3. A sheep or a goat

THE OBLIGATORY ELEMENTS OF HAJJ

Ibn 'Ashir mentions eleven obligations of hajj:

1. Going around the Ka'ba on arrival (*tawaf al-Qudum*)
232 2. Immediate performance of *sa'y* thereafter – but if these two are left out because of forgetfulness or because one fears one will not make it to 'Arafa on time, no *hady* is required
3. Doing both *sa'y* and *tawaf* on foot if able
4. Doing two rak'as after any circumambulation that is obligatory or a pillar. If one abandons these two and gets far from Makka, one must

sacrifice an animal.

5. Stopping over at Muzdalifa after returning from 'Arafa (staying until Fajr is sunna) **233**
6. Spending three nights after 'Arafa in Mina
7. Intending the state of *ihram* when passing (by air, land or sea) the designated entry points which are of two kinds: **234**
   - □ Entry points related to space (*miqat al-makani*), which are also of two kinds:
     - The traveller intending Makka has the following *miqat*:
       - Dhu'l-Hulayfah if coming from the direction of Madina (North)
       - Al-Juhfah if coming from the direction of Egypt or Sham i.e. Lebanon, Palestine, Syria and Jordan (North-West) **235**
       - Qarn if coming from the direction of the Najd (East)
       - Dhatu 'Irq if coming from the direction of Iraq (North-East)
       - Yalamlam if coming from the direction of Yemen (South)
     - For the permanent or temporary resident[185] of Makka, it is *mandub* that he intends *ihram* from the *Masjid al-Haram*. For the temporary resident only, it is *mustahabb* to go out to his *miqat* – if there is time. (See section on 'umra for related details)
   - □ Entry points related to time (*miqat az-zamani*): 1st of Shawwal to Fajr of the day of 'Id for hajj (and the whole year round for 'umra)
8. For men not to wear sewn clothes or sewn footgear beyond the *miqat*. Men wear slippers and two white cotton sheets – towelling is better against both heat and cold – and women wear their normal clothes – see section on *ihram*. **236**
9. Saying the *talbiyya* (moderately loudly for men, quietly for women): "*Labbayka. Allahumma labbayk. Labbayka. La sharika laka labbayk. Inna-l-hamda wa-n-ni'mata laka wa-l-mulk. La sharika lak.*" ('At Your service. O Allah, at Your service. At Your service, You have no partner, at Your service. Truly, the praise, blessing and kingdom are Yours. You have no partner.')
10. Shaving or cutting one's hair after throwing stones at the biggest monolith the first time at Mina
11. Throwing stones at the monoliths on the days of Mina

185 There is no category of 'traveller' in Makka related to hajj and 'umra, only for *salat*.

DESCRIPTION OF HAJJ

237 This combines the elements of hajj in their correct order.

Assuming ihram

238 1. When reaching the *miqat*[186]
2. One puts into practice all the elements of the *fitra*[187]
3. Performs a ghusl (if in *janaba* one must intend to lift this condition simultaneously)[188]

239 4. Puts on the clothes of *ihram*, consisting on a loin-cloth (*izar*), a shawl (*rida'*) and a pair of sandals (*na'layn*)
5. Brings a sacrificial animal if one has one and it is possible to do so
6. And prays two rak'as

240 7. Reciting respectively surat al-Kafirun and al-Ikhlas after al-Fatiha
8. If travelling overland, one mounts one's vehicle or beast and

241 9. Makes the intention to enter *ihram* for hajj, immediately setting off from the *miqat*, and uttering the *talbiyya*[189]

242 10. And renewing the *talbiyya* whenever one's condition changes, such as:
    - ☐ Standing up, sitting or lying down
    - ☐ Going up or down a hill
    - ☐ Mounting or dismounting a beast or vehicle
    - ☐ Meeting another group of *hajjis*
    - ☐ When hearing other people saying it
    - ☐ Waking from sleep
    - ☐ And immediately after performing salat

    – however it is *makruh* to say the *talbiyya* non-stop.

186 At Rabigh, because it belongs to al-Juhfa.

187 By shaving the pubic hair, removing the hair under the armpits, trimming the moustache and clipping the nails.

188 This *ghusl* is for adults, children, menstruating woman and the one affected by afterbirth bleeding. This ghusl is done rubbing and removing the dirt, contrary to the other ghusls of hajj where the hand just passes over the body without removing any dirt.

189 The *talbiyya* should be done at a moderate volume, not too loud or too low.

Reaching Makka

11. If possible one performs a *mustahabb* ghusl (without rubbing the skin – see section on *ihram*) at Dhu Tuwa when approaching Makka. **243**
12. Entering by Kuda ath-Thaniyya
13. When the boundaries of the city of Makka are reached, the *talbiyya* is stopped and all occupation with worldly affairs is left as one proceeds towards... **244**
14. ...The Masjid al-Haram as quickly as possible through the gate called Bab as-Salam. **245**

Tawaf al-Qudum

15. One proceeds directly to greet the Black Stone as described in point 21, without performing two rak'as to greet the Mosque. This is the first action of *tawaf al-Qudum*. (One must intend *tawaf al-Qudum* for hajj or *tawaf al-'umra* if performing 'umra).
16. Now say the *takbir* ('Allahu akbar' – Allah is the Greatest).
17. Begin the seven-round circumambulation called *tawaf al-Qudum*, keeping the Ka'ba on the left hand side. **246**
18. Repeat the *takbir*.
19. And the greeting of the Black Stone each time it is passed as described in point 21. **247**
20. Also say the *takbir* at the Yamani corner. Touch it if possible and bring the hand to the mouth, but neither the hand nor the Yamani corner is kissed.
21. When greeting the Black Stone one first tries: **248**
    - ☐ Kissing it, and if this is impossible because of crowds, then
    - ☐ Touching it and bringing one's hand to one's mouth without actually kissing the hand
22. Do the first three rounds trotting (men only)[190] and the last four walking. **249**
23. Then, pray two rak'as behind Maqam Ibrahim (without preventing people from passing right in front) reciting respectively surat al-Kafirun and surat al-Ikhlas after al-Fatiha.
24. Then supplicate[191] freely at the Multazam area between the Black Stone and the door of the Ka'ba. **250**

---

190 This trotting is only done by men and only in the first *tawaf* (*Tawaf al-Qudum*).

191 This is recommended.

Sa'y

25. Greet the Black Stone as the first of the sunna of *sa'y* and
251 26. Go to Safa, look towards the Ka'ba and say:
   - □ *Allahu akbar* (3 times) *la ilaha illa'llahu wahdahu la sharika lah, lahu'l-mulku wa lahu'l-hamd, wa huwa 'ala kulli shay'in qadir. La ilaha illa'llahu wahdah, anjaza wa'dah, wa nasara 'abdah, wa hazama'l-ahzaba wahdah.* (Allah is the Greatest (3 times), there is no god except Allah alone without partner, His is the kingdom and His is the praise, and He has power over everything. There is no god except Allah alone, He fulfilled His promise, and helped His slave, and He alone defeated the Confederates).
   - □ Supplicate freely.
   - □ Do *salatu 'ala'n-nabiyy.*
252 27. And then go to Marwa (and do what is mentioned in point 26 above).
28. Go with haste in the section called Batn al-Masil[192] as it is sunna.
253 29. Stop four times at Safa and four times at Marwa (ending up at Marwa) seven runs in total.
254 30. One supplicates freely during both *sa'y* and *tawaf* with humbleness and admission of one's shortcomings.

*Rulings of **tawaf** and **sa'y***

255 31. The rulings of the *tawaf* are divided into two categories:
   - □ The obligatory elements, which are eight:
     - Being free of impurity in body, clothes and place
     - Being in a state of wudu' (or tayammum if necessary)
     - Covering the *'awra*
     - Performing seven circumambulations
     - And performing them consecutively
     - Inside the limits of the Mosque and
     - Outside the rim at the bottom of the Ka'ba called *ash-shadarwan* and the tomb of Isma'il called al-Hijr
     - And to keep the Ka'ba on the left, i.e. performing the *tawaf* counter-clockwise

192 Nowadays this section is marked by two green lights.

- ☐ The sunna elements which are divided into four categories:
  - To do it on foot if able
  - To kiss The Black Stone and to touch the Yamani Corner at the beginning
  - Du'a and *salatu 'ala an-nabiyy*
  - Men should trot in the first three rounds of *tawaf al-Qudum* only.

The rulings of the *sa'y* are divided into three categories:

- ☐ The obligatory elements, which are three:
  - Seven runs
  - Starting from Safa
  - Preceded by a valid *tawaf*
- ☐ The sunna elements, which are four:
  - Kissing the Stone after the two rak'as of *tawaf* at Maqam Ibrahim
  - Climbing the hills, Safa and Marwa
  - For men to hasten between the two hills
  - To make du'a
- ☐ The *mustahabb* elements, which are three:
  - Being free of impurity in body, clothes and place
  - Being in a state of wudu' (or tayammum if necessary) and
  - Covering the *'awra*

Waiting for 'Arafa

32. The *talbiyya* is now resumed, and continued until the time of Dhuhr **256**
on the 9th of Dhu'l-Hijja.
33. At the Masjid al-Haram, after *salat adh-Dhuhr* on the 7th of Dhu'l-Hijja,[193] it is recommended to attend the khutba where the rites of hajj are described by the Imam.[194]
34. Then, on the 8th[195] one leaves Makka to arrive at Mina at the time of **257**

193 Called *Yawm az-Zinah.*

194 This khutba is given by the imam after the salat and there is no sitting in the middle. The imam describes what is to be done from that day until noon on the Day of 'Arafa.

195 Called *Yawm at-Tarwiya.*

Dhuhr, where the salats are shortened, but not joined. The pilgrim stays in Mina for the rest of the day and the night until, he prays Subh.

'Arafa

35. On the 9th, after sunrise, one sets out to arrive at the plain of 'Arafa.

258 36. There, a *mustahabb* ghusl is performed (without rubbing the skin) shortly before the time of Dhuhr.

37. And at noon the *talbiyya* is stopped for good.

38. It is *mustahabb* to head for the Mosque of Namira, to listen to the two khutbas, where the imam describes the rites to be done until the 11th of Dhu'l-Hijja.

39. The salats of Dhuhr and 'Asr are shortened and joined, whether they are prayed with the Imam at the Namira mosque, done in another *jama'ah* or by oneself.

259 40. Then, in a state of wudu', the foot of the Mount of Mercy (*Jabal ar-Rahma*) is ascended. That is the preferred place, but the whole of the plain or 'Arafa is valid. One stands, preferably mounted on a beast, because that is the Sunna, if not then standing is better than sitting. One should not sit unless tired. As for women, it is better to sit.

41. Do abundant, fervent and constant dhikr consisting of:

260 □ Supplication

□ Uttering "*la ilaha illa'llahu wahdahu la sharika lah, lahu'l-mulku wa lahu'l-hamd, wa huwa 'ala kulli shay'in qadir*"

□ And *salatu 'ala'n-nabiyy*

This is done facing the qibla until sunset.

261 42. After sunset, one waits a moment within the boundaries of 'Arafa[196]

Muzdalifa

43. Then one sets out for Muzdalifa.

262 44. Passing between the two mountains of Ma'zamain which are landmarks on the route, doing dhikr.

196 To stay in 'Arafa a part of the night is a pillar of the hajj. Staying in 'Arafa until sunset fulfils that pillar.

45. And upon arrival at Muzdalifa, one joins *salat al-Maghrib* and *salat al-'Isha* (shortening *salat al-'Isha* to two rak'as) – with the imam if one is able.
46. One postpones eating and unloading heavy luggage until after the salat. **263**
47. Keeping the night alive with worship (salat, dhikr and du'a) is *mustahabb*.
48. In the morning, on the day of 'Id (10th of Dhu'l-Hijja), it is *mustahabb* to pray *salat as-Subh* in Muzdalifa and to set out while it is still dark.
49. Move to the area called Mash'ar al-Haram and do much du'a (facing qibla and with the Mash'ar to one's left) until just before sunrise. **264**
50. Before leaving, one picks up seven bean-sized stones for stoning the *jamarat* of 'Aqaba at Mina (the rest of the stones may be picked up outside Muzdalifa).

First stoning at Mina

51. Then one hastens through the Valley of Fire (*Wadi an-Nar*), returning to Mina.
52. To pelt the *jamarat* of 'Aqaba (the monolith closest to Makka) **265**
53. Hitting it towards its base, using the seven stones picked up at Muzdalifa and paying attention to the following rulings: **266**
    - □ The stones should be bean-sized (smaller stones are invalid, bigger are *makruh*).
    - □ It is *makruh* to use stones previously used for stoning.
    - □ It is invalid to throw pieces of metal or clay.
    - □ It is *mandub* to say *takbir* with each throw.
54. And when this is done the limitations of *ihram* are lifted except three:
    - □ Sexual relations remain haram.
    - □ Hunting remains haram.
    - □ Perfume remains *makruh* (also see section on *ihram*).
55. Now the sacrificial animal may be slaughtered and two conditions apply:
    - □ If one brought it into the sacred area from 'Arafa it is slaughtered at Mina. **267**
    - □ But if not, it is slaughtered later in Makka after being brought into the Sacred Precinct from another location.

56. And finally one either shaves or cuts one's hair (doing either is obligatory). Shaving is preferred for men, but if cutting, the remaining hair length must be short, 1-2 cm left at most. Cutting is preferred for women, even if it is just the length of a third of a finger

Tawaf al-Ifada

57. Return to Makka to perform *tawaf al-Ifada* as described previously and to pray two rak'as, preferably wearing the clothes of *ihram* –and if one did not do *sa'y*, it must be done now. By this the limitations of *ihram* are lifted completely.

Second stoning at Mina

268 58. Now return to Mina on the 10th of Dhu'l-Hijja.
59. And arriving in time to pray Dhuhr there is *mustahabb.*
60. Spending most of each of the nights in Mina is obligatory.
61. The day after the 'Id (11th of Dhu'l-Hijja), during the afternoon, but before praying Dhuhr, stone all three *jamarat*, with the following conditions applying:
    - ☐ Go to them in a state of wudu and
    - 269 ☐ Using the 21 stones that one brought along, seven per monolith
    - ☐ Throwing them one at a time, and
    - ☐ After the time of Dhuhr, or the stoning is invalidated
    - 270 ☐ Move forward and stop to make du'a facing the qibla after the first and second monolith, but not after the last one
    - ☐ Starting with the monolith furthest from Makka, ending with the closest one (*Jamarat al-'Aqaba*)
    - ☐ Uttering *takbir* with each throw and
    - ☐ Waiting the time it would take to recite Surat al-Baqarah at a quick pace (about half an hour) between each monolith

271 62. The next day (12th of Dhu'l-Hijja), repeat the stoning as described above.
63. On the 13th, one has two choices, depending on one's desire – one may:
    - ☐ Leave Mina, which ends the hajj (but one must exit the limits of Mina before the adhan of Maghrib or it becomes

obligatory to stay the night and repeat the stoning the next day) or

- □ One may stay the night and repeat the stoning again the next day at which point the hajj ends with the last stone thrown.

IHRAM

Ihram is the state resulting from the intention to go on hajj and this state makes the following haram: 272

1. Hunting[197]
   - □ Any land animal (i.e. sea animals are not included in this prohibition)
   - □ Birds (i.e. excluding chickens)
   - □ Or the offspring of these land animals and birds (including breaking their eggs)
   - □ And parasites including lice, fleas or mites

– all demand the ransom (*jaza'*) described below, except killing six sorts:

- Rats and mice 273
- Scorpions
- Kites
- Snakes
- Crows
- And any carnivores, including dogs, which pose a threat

– Killing animals during *ihram* demands compensation (*jaza'*) and one may choose between:

- □ Slaughtering an animal – the type of which is decided by two trustworthy *fuqaha*[198] to correspond to the animal killed – either:
  - A sheep or goat or

197 Even disturbing these animals – by scaring, hurting, ensnaring, trapping or throwing things at them or such – is haram, but compensational sacrifice is only performed if the animal is killed. This ruling covers the person in *ihram* wherever he is, but applies to everyone inside the Sacred Precinct whether in *ihram* or not.

198 *Fuqaha* (plural of *faqih* – expert in fiqh).

- A cow or
- A camel or

□ Feeding poor people (preferably who live close to where the animal was killed) as many *mudd* of food as the *fuqaha* decide or

□ Fasting one day per *mudd*

274 2. Wearing certain types of clothing depending on gender:

□ For men there are two restrictions; they must not:

- Wear clothes which are stitched or closed by buttons, Velcro, zippers, indeed anything that forms a closed loop, such as:

- the neck and sleeves of jallabas and
- shirts
- capes closed in the neck
- trousers
- rings
- bracelets (including watches)
- *khuffs*
- gloves
- belts – except with the intention to carry money
- and sleeping in a zipped sleeping bag and

275 • Wear anything like hats, turbans, etc. which cover the face or head

276 □ As for women, they are only prevented from two things:

- Wearing gloves and
- Covering the face – except a facial cover (which hangs from above and does not stick to the skin of the face) adopted for reasons of modesty

277 3. Applying perfume – smelling it is *makruh*, and as for using it, it is:

□ Haram if it is oud, musk, amber, camphor, or saffron, etc. [199]

□ But only *makruh* if it is the perfume of flowers, like rose, jasmine and the like[200]

4. Ointments and creams, except non-perfumed moisturising cream to

199 These are called *female* perfumes because when applying them some of its substance sticks on the body. Included in this category are all modern perfumes, colognes, oils and ointments.
200 These are called *male* perfume.

avoid cracking of hands and soles of feet[201]

5. Enjoying the comforts of personal hygiene that remove – by washing, rubbing or cutting – certain items from oneself or others:
   - Lice
   - Dead skin
   - Dirt (except ritual impurity)
   - Nails and
   - Hair

   Ransom (*fidya*) must be paid for violating points 2 to 5, even if one had a valid excuse. One may choose freely between: **278**

   - Slaughtering a sheep, goat, cow or camel or
   - Feeding six poor people two *mudds* of grain or
   - Fasting three days (need not be consecutive)
6. Also three types of relation between the sexes are haram: **279**
   - Marriage or marriage contracts and
   - Preliminaries to sex (from winking to touching), which demand *hady*
   - But penetration, orgasm or ejaculation (by any means except a wet dream) nullifies the hajj.

Coming out of *ihram* (*tahallul*) is in two stages: **280**

1. The minor *tahallul*, which precedes the major and takes place with the stoning of *jamrat al-'Aqaba*. All prohibitions become permissible except no. 1 (hunting) and no. 6 (sexual relations).[202]
2. The major *tahallul*, where one leaves *ihram* completely and the rest of the restrictions are lifted, is not until the *tawaf al-Ifada*[203] is performed on the 10th of Dhu'l-Hijja.[204] If penetration, orgasm or ejaculation occurs before the major *tahallul*, the hajj or 'umra[205] is nullified and it is obligatory to:

201 Using perfumed shampoo and soap is *makruh*.

202 Although the use of perfume remains *makruh* until the major *tahallul* takes place.

203 The *tahallul* happens and the restrictions are lifted for those who performed the *sa'y* before 'Arafa. If not the *tahallul* will only take place after the performance of the *sa'y* after the *tawaf al-Ifada*.

204 And this is on the condition that the pilgrim shaves his head; if not, sex is not permitted and he will have to slaughter the *hady* if he or she has sex.

205 The 'umra is nullified if sexual intercourse occurs before finishing the *sa'y*. If it occurs after

- ☐ Continue the hajj until the end
- ☐ Make up the hajj next year (even if the nullified hajj was not obligatory)
- ☐ Slaughter the *hady* then – if one cannot slaughter, one must fast ten days

281 Regarding seeking shade, there are two rulings depending on the object:

1. As for mobile objects, i.e. that are not fixed in the ground, such as umbrellas, it is permissible to seek shade next to them, but not under them, and there is disagreement whether *fidya* is obligatory or *mustahabb.*
2. As for fixed objects, including raised tents, both to seek shade next to them and under them are permissible.

ʿUMRA

282 Performing ʿumra[206] once in a lifetime is a sunna *muʾakkada*. The ʿumra can be performed throughout the year, but it is *makruh* to do it more than once a year. Its rites of *ihram*, *tawaf* and *saʿy* are performed like those of hajj.

It is *mustahabb* to assume *ihram* at Tanʿim if coming from inside Makka, but if coming from outside, the entry points (*miqat*) will be the ones described before for those who perform hajj.

283 *ʿUmra* ends with cutting the hair after the *saʿy* and this lifts all restrictions and completes it.

ADAB OF THE HARAM OF MAKKA

While in Makka one should:

1. Do much *tawaf*, especially those who are not from Makka.
2. 284 Be mindful of the sanctity of the place and its surroundings by avoiding obscenity, transgression and dis-obedience, and do many acts of obedience and worship.

---

the *saʿy* but before shaving the hair, it is not nullified, but the *hady* will have to be sacrificed.
206 The rites of ʿumra consist of *ihram*, *tawaf* and *saʿy*.

3. Make sure to pray in the communal *fard* salat, as the reward for praying at Allah's House is greater than in any other mosque except the Prophet's ﷺ Mosque in Madina, according to the Malikis. 285
4. It is recommended to perform the farewell *tawaf* (*tawaf al-Wada'*) before leaving.[207]

VISITING THE TOMB OF THE PROPHET ﷺ

Visiting the Chosen One is a sunna, and is done with great *adab* and clear intention so that one will be answered in all of one's du'as to Allah. Its *adab* is the following: 286

1. To do a lot of *salatu 'ala an-nabiyy* at all times.
2. Whilst on the way to Madina, to make *takbir* when going up hills.
3. To stop outside Madina in order to do ghusl, pray two rak'as, put on one's best clothes, apply perfume and renew one's *tawba.*
4. To walk on foot towards the mosque.
5. Upon arrival at his Mosque, to start by praying two rak'as, if it is at a time where one may pray.
6. Then, proceed to stand in front of his tomb facing it (but without clinging to it) and greeting him ﷺ. The person doing so should do it with humility and feel that he is standing in front of the Messenger of Allah ﷺ. 287
   - ☐ The greeting is: "*as-Salamu alayka ayyuha'n-nabbiyu wa rahmatullahi wa barakatu. Sall'Allahu 'alayka wa 'ala azwajika wa dhurriyatika wa 'ala ahlika ajma'in. Fa qad ballaghta'r-risala, wa addayta'l-amana, wa 'abadta rabbak, wa jahadta fi sabilih, wa nasahta li 'abidih, sabiran muhtasiban hatta ataka'l-yaqin.*"[208]
   - ☐ Then, proceed to greet Sayyiduna Abu Bakr as-Siddiq to the right saying: "*as-Salamu alaykum ya Aba Bakrin is-Siddiq wa rahmatullahi wa barakatuh. Safiyya rasulillahi (sall'Allahu 'alayhi wa sallam) wa thanihi fi'l-ghar. Jazak'Allahu 'an ummati*

207 This *tawaf* is performed in the way already described in the rites of hajj.

208 Peace be upon you, Prophet, and the mercy and blessings of Allah. May Allah bless you, your wives, your descendants and all your family. You have, certainly, delivered the message, fulfilled the responsibility, worshiped your Lord, fought in His way and advised His servants with patience and devotion until you reached certainty.

rasulillahi ﷺ.*"[209]

- Finally, proceed further to the right to greet 'Umar ibn al-Khattab saying: "*as-Salamu alaykum ya Aba Hafsin al-Faruq wa rahmatullahi wa barakatuh. Jazak'Allahu 'an ummati Muhammad* ﷺ *khayra.*"[210]

288 7. Then supplicate to Allah at his tomb ﷺ for three things, knowing that in this place all your requests will be answered:
   - The best in dunya and akhira
289 - The intercession of the Prophet ﷺ and
   - A good seal on your life, i.e. death while declaring the Shahada
8. And perform a lot of *nafila* rak'as in the area known as ar-Rawda between his tomb and his *minbar.*

RETURNING HOME

When returning home it is recommended to:

1. Do so quickly when finished.
290 2. Arrive at the time of *duha* (i.e. when the sun is up in the sky).
3. Bring gifts to relatives and friends if able.

209 Peace be upon you, Abu Bakr as-Siddiq, and the mercy and blessings of Allah. The best friend of the Messenger of Allah ﷺ and the second with him in the cave. May Allah give you a good reward on behalf of the Ummah of the Messenger of Allah ﷺ.

210 Peace be upon you, Abu Hafs al-Faruq, and the mercy and blessings of Allah. May Allah give you a good reward on behalf of the Ummah of Muhammad ﷺ.

The Principles of Tasawwuf
and the Guide to Gnosis

The Principles of Tasawwuf and the Guide to Gnosis

INTRODUCTION TO TASAWWUF

TASAWWUF IS THE science of purifying the heart and character in order to attain nearness to Allah. The heart is the seat of intention – and all action is based on intention, as is known, and thus tasawwuf is the heart of Islam.

Imam Malik said: "Whoever takes on tasawwuf without taking fiqh becomes heretical and whoever takes on fiqh without taking tasawwuf becomes a transgressor. Only he who combines the two will attain to the truth."[211]

Ibn Khaldun says that this purification "...was the general rule among the [...] *salaf*, but when involvement in dunya became widespread from the 2nd Century [...] those devoted to worship came to be called *sufiyya* or *ahl at-tasawwuf*."[212]

It is said that the name is from *suf* (wool) which these people wore. Or that it refers to *ahl as-suffa*, the poor Companions who left everything to live on the bench (*suffa*) by the Mosque of the Prophet ﷺ. Other things are also said as far its etymology is concerned, and Allah knows best.

Tasawwuf is not taken from books, as inner purity is not quantifiable like fiqh (i.e. the knowledge that Dhuhr is four rak'as). Rather, it is a knowledge passed on directly from person to person in an unbroken chain back to the Prophet ﷺ and the heart is its container.[213]

---

211 *Iqadh al-Himam fi Sharh al Hikam*. Al-Maktaba al-Thaqafiya, p. 5.

212 From: Ibn Khaldun, 'Abd ar-Rahman. *Muqaddima Ibn Khaldun*. Reprint. Mecca: Dar al-Baz, n.d.

213 The people of tasawwuf refer to the knowledge and its container, in a metaphorical way, as to the wine and the cup.

The teachings passed down in such a chain is called a *tariqa* (path) and, while the aim is identical, the methods may vary from one *tariqa* to another, and within each depending on time and place.

Many *tariqas* trace back to Imam al-Junayd (d. 297), who was the first to formalise tasawwuf as a science. He said: "All the paths to Allah are closed save for the one travelling in the footsteps of Muhammad." Or, in other words, there is no tasawwuf outside of the Shari'a.

TAWBA

291 The first element of tasawwuf is that *tawba* is obligatory immediately and absolutely[214], and any postponement demands a separate *tawba* for the postponement.

Tawba means to turn away from all transgressions (*ma'siya*) with sincere regret. Transgressions can be:

- minor (*saghira*)
- major (*kabira*)
- against Allah
- against people
- known
- unknown

292 *Tawba* has three conditions:

1. Ceasing the act (*iqla'*); such as stopping if drinking wine; or cutting off one's words in mid-sentence when backbiting
2. Sincerely intending never to repeat the transgression (*nafy al-israr*)
3. And repairing damage caused to people or property as best as one is able (*talafi al-huquq*); such as refunding stolen goods

Asking for forgiveness in abundance is desirable (*mandub*), by uttering the words "*astaghfirullah*" ("I seek forgiveness from Allah").

Tawba is the first step to *taqwa*.

214 It is obligatory to do *Tawba* from all kind of transgressions.

Tawba is obligatory because of the following evidence:

1. From Allah's Book:
   - ☐ *Make tawba to Allah, every one of you, muminun, so perhaps you will have success.* (24:31)
   - ☐ *Ask forgiveness of your Lord and then make tawba to Him.* (11:52)
   - ☐ *You who have Iman! Make sincere tawba to Allah. It may be that your Lord will erase your wrong actions and admit you into Gardens with rivers flowing under them.* (66:8)
   - ☐ *Certainly Allah loves those who do tawba and He loves those who purify themselves.* (2:222)
2. From His Messenger ﷺ who said: "O People! Make *tawba* to Allah and ask His forgiveness, for I certainly make *tawba* to Allah a hundred times a day." (*Sahih Muslim*)

TAQWA

Taqwa is to protect one's self (*nafs*) from that which will harm one in the akhira, and it is obtained by four means: **293**

1. Performing every outward obligation
2. Performing every inward obligation
3. Avoiding every outward haram
4. Avoiding every inward haram

In these four means, there is benefit for the spiritual traveller (*as-salik*) i.e. the *murid*. **294**

The following fifteen benefits of taqwa are taken from the exegesis (*tafsir*) of the Qur'an by Imam Ibn Juzayy:[215]

1. Guidance, because of His words: *"...guidance for the people of taqwa"* (2:2)
2. Help because of His words: *"Truly, Allah is with the people who have taqwa"* (2:194)
3. Close friendship because of His words: *"Allah is the close friend of the people of taqwa"* (45:19)
4. Love because of His words: *"Truly Allah loves the people of taqwa"* (3:76)

215 From *at-Tashil li 'Ulum at-Tanzil.*

5. Covering of wrong actions because of His words: *"If you have taqwa of Allah He will give you a discrimination and He will cover your wrong actions"* (8:29)
6. A way out from unhappiness, and
7. Provision from where one does not expect because of His words, *"Whoever has taqwa of Allah He will make a way out for him and provide him from where he does not expect"* (65:2-3)
8. Facilitation of affairs because of His words: *"Whoever has taqwa of Allah He will make ease for him in his affair"* (65:4)
9. Full covering of wrong actions and
10. Magnification of rewards because of His words: *"Whoever has taqwa of Allah He will cover over his wrong actions and magnify a reward for him"* (65:5)
11. Acceptance of actions because of His words: *"Allah only accepts from the people of taqwa"* (5:27)
12. Success because of His words: *"Have taqwa of Allah in order that you might succeed"* (2:189)
13. Good news because of His words: *"Those who have Iman and taqwa, for them there is good news in this world and in the next"* (10:63-63)
14. Entrance into the Garden because of His words: *"Truly, there are for the people of taqwa with their Lord Gardens of bliss"* (68:34)
15. Salvation from the Fire because of His words: *"Then We will save the ones who had taqwa"* (19:72)

Taqwa is obligatory because of the following evidence:

1. From Allah's Book:
   - ☐ *"The noblest among you in Allah's sight is the one with most taqwa"* (49:13) and
   - ☐ *"You who have Iman! Have taqwa of Allah with the taqwa due to Him"* (3:102) and
2. From His Messenger ﷺ:
   - ☐ "The first thing that will make a person enter *Jannah* is *taqwa* of Allah and good character." (*Sunan* at-Tirmidhi, declared *sahih* by al-Hakim)

THE OUTWARD HARAM

295 Avoiding what is haram is harder on the *nafs* than performing what is

obligatory. The soundness or corruption of the actions of the limbs originate in the heart, as it is their master.

The Messenger of Allah ﷺ said:

"Is there not in the body a piece of flesh, that if sound the body is sound, if corrupt, the body is corrupt? Is it not the heart?" (*Sahih* Muslim and *Sahih* al-Bukhari)

As for the outward haram mentioned, one must guard the seven parts of the body.

The Messenger of Allah ﷺ said:

"Allah has created seven doors to the Fire, and He has created in the son of Adam seven parts of the body; whenever he obeys Allah with one of these seven parts, Allah closes one of the doors of the Fire, and whenever he disobeys Allah with one of these seven parts, it becomes incumbent on him to enter through it." (*The Musnad* of Ahmad ibn Hanbal)

Guarding the seven parts of the body is done by:

1. Lowering the eyes from what it is haram to look at, such as:
   - ☐ People's nakedness, including any form of image of nakedness
   - ☐ People's physical defects
   - ☐ Women (or children) with lust
   - ☐ Snooping regarding people's property
   - ☐ Looking at other Muslims with contempt
2. Shutting the ears from offensive speech, such as:
   - ☐ Backbiting (*ghiba*), i.e. to mention anything behind a person's back that he would dislike 296
   - ☐ Tale-carrying (*namima*), i.e. to pass on statements which damage relations between people
   - ☐ Perjury (*zur*), i.e. to give false testimony in court to a judge (*qadi*)
   - ☐ Lying (*kadhib*) i.e. to knowingly claim that something is otherwise than what it really is
   - ☐ Listening to the voice of an unrelated member of the opposite sex with lust

3. And guarding one's tongue from uttering the above, especially obscene language

297 4. Guarding the stomach from the haram, including:
   - □ Food which has been usurped or
   - □ Bought with money which has been stolen, usurped or acquired by usury
   - □ Unslaughtered meat[216]
   - □ Flowing blood
   - □ Pork
   - □ Intoxicating beverages, such as wine etc.
   - □ Drugs, including hashish

   The haram is not restricted only to the stomach but applies to everything we use, wear, mount or dwell in. Add to that, if any of these things are acquired by illegal means such as usury, theft or usurpation it will also be unlawful to use. And Allah knows best.

   Intentionally abandoning what is doubtful (i.e. that about which there are differences of opinion) yields a reward in the akhira, in all matters, including dwelling, clothes, utensils and such.

298 5. Guarding the private parts from sodomy, fornication and adultery – and masturbation, according to the strongest position
6. And finally fearing Allah, the ever-present Witness, and not letting the hands reach for what is haram and
7. Not letting the feet carry one to what is haram

One should not undertake any matter until one knows Allah's judgement on it, by asking the people of knowledge. This means that it is obligatory for a person involved in selling, renting, business partnership, etc. to know the general rules pertaining to these.

216 There is a difference of opinion as to whether meat slaughtered by the People of the Book is halal if slaughtered differently to the Muslim way. Qadi Abu Bakr ibn al-'Arabi mentions in *Ahkam al-Qur'an* that the meat slaughtered by Christians is halal for Muslims if it is acceptable to their priests, even if it is slaughtered differently from the Muslim way. This view is also mentioned in *al -Mi'yar al-Mu'rib* by al-Wansharisi.

THE DISEASES OF THE HEART

The inward haram states are called the diseases of the heart, and it is obligatory to dispel them.[217] They include:

1. Showing off (*riya'*), which is seeking position in the hearts of people by showing good characteristics. Its signs are laziness and few actions when alone and lots of activity around people, especially when one is praised. It is also known as 'the small *shirk*'. **300**

   Showing off is haram because of the following evidence:

   - From Allah's Book: "*Woe to those who* pray, *and are forgetful of their salat, those who show off...*" (107:4-6) and
   - From His Messenger ﷺ:
     - The *hadith qudsi*: "Allah says: 'Whoever does an action for Me, in which he associates in it other-than-Me, the action is all for him and I am free from it, and I am the most independent of those independent of partners.'" (Muslim, from Abu Hurayra) and
     - His saying, "The thing that I most fear for you is the small *shirk*". They said, "What is the small *shirk*?" He said, "Showing off. Allah will say on the Day of Rising, when He will reward His servants for their actions, 'Go to those to whom you were showing off in the dunya, and see whether you find any reward with them.'" (Ahmad, Ibn Maja and al-Hakim, from Mahmud ibn Labid)
2. Love of prestige (*sum'a*), which is wanting people to speak well about oneself, and is related to showing off
3. Envy, which subdivides into two:
   - *Hasad* – wanting people to lose what Allah has given them, even without wanting what they have for oneself – is haram, except against a transgressor or a disbeliever who is using the gifts Allah has given him in the wrong way.
   - *Ghibta* – wanting the same as other people have, without wanting them to lose it. It can be obligatory, as when wanting

217 Imam al-Ghazali said that to cure the heart from its spiritual diseases is a personal obligation (*fard 'ayn*).

to have other people's resolution to perform obligatory actions. It can also be recommended or permitted.

Envy (*hasad*) is haram because of the following evidence:

- ☐ From Allah's Book:
  - About the people He has cursed *"...do they in fact envy other people for what Allah has given them?"* (4:53)
  - And about the enemies of His Messenger ﷺ: *"If something good happens to you, it galls them."* (3:120)
- ☐ From His Messenger ﷺ: "Beware of envy because envy consumes good actions as fire consumes wood." (*Sunan* Abu Dawud and ibn al-Majah, from Abu Hurayra)

4. Conceit (*'ujb*), which is the glorification of one's good actions while forgetting they are a gift from Allah

   Conceit is haram because of the following evidence:

   - ☐ From Allah's Book:
     - *"..When your great numbers delighted you, but did not help you in any way."* (9:25) and
     - *"..They thought their fortresses would protect them. Then Allah came upon them from where they least expected it and cast terror into their hearts"* (59:2) and
     - *"... Do not claim purity for yourselves."* (53:31)
   - ☐ From His Messenger ﷺ: "Three things are destructive, greed obeyed, whims followed and self-admiration." (Al-Bazzar, at-Tabarani, al-Bayhaqi, Abu Nu'aym from Anas ibn Malik)
   - ☐ And from his companions, Ibn Masud ؓ said: "Destruction lies in two things, despair and conceit" – because the one who despairs does not seek goodness, and the one who is conceited thinks he already has it."

5. Arrogance (*kibr*) which is the glorification of oneself, while despising other people

Other diseases briefly mentioned by Mayyara include:

6. Rancour (*ghill*)
7. Hatred (*hiqd*)
8. Rebelliousness (*baghy*)
9. Anger for the sake of other-than-Allah (*ghadab li ghayrillah*)

10. Fraud (*ghish*)
11. Stinginess (*bukhl*)
12. Denying the truth out of arrogance (*al-i'rad 'ani l-haqq istikbaran*)
13. Involvement in that which does not concern one (*al-khawd fi ma la ya'ni*)
14. Avidity (*tam'*)
15. Fear of poverty (*khawf al-faqr*)
16. Displeasure with fate (*sakhat al-maqdur*)
17. Insolence (*batar*)
18. Praising the rich for their wealth (*ta'dhim al-aghniya li ghinahum*)
19. Despising the poor for their poverty (*istihza' bi'l-fuqara' li faqrihim*)
20. Pride (*fakhr*)
21. Presumptuousness (*khayla'*)
22. Competing for dunya (*at-tanafus fi'd-dunya*)
23. Boasting (*mubaha*)
24. Self-embellishment for the sake of creatures (*at-tazayyun li-l-makhluqin*)
25. Flattery (*mudahana*)
26. Love of praise for what one does not do (*hubb al-madh bi ma la yaf'al*)
27. Looking at other peoples defects (*al-ishtighal bi 'uyub an-nas 'an 'uyibih*)
28. Ingratitude (*nisyan an-ni'ma*)
29. Fanaticism (*hamiyya*)
30. And fearing and hoping for other-than-Allah (*ar-raghba wa ar-rahba lighayrillah*)
31. Loving leadership (*hubb ar-ri'asa*) – this is the last disease to be removed from the heart of the *siddiqun* **301**
32. Forgetting the hereafter (*nisyan al-akhira*) and entertaining false hopes
33. Love of dunya (*hubb ad-dunya*), which is at the head of all wrong action according to the Messenger of Allah ﷺ **302**

Forgetting death and love of dunya are haram because of the following evidence:

- □ From Allah's Book: *"No! But you love the immediate and leave the Hereafter"* (75:20-21)
- □ From His Messenger ﷺ:
  - "Certainly the dunya is cursed, and cursed is what is

in it, except the dhikr of Allah, and what He loves: someone who possesses sacred knowledge, and someone learning it." (At-Tirmidhi and Ibn Majah from Abu Hurayra)

- "If dunya had the value with Allah of the wing of a gnat, He would never have given the *kafir* a sip of water." (At-Tirmidhi from Sahl ibn Sa'd)
- "Be in the dunya as a stranger or traveller." (Al-Bukhari, from Ibn 'Umar)
- "Do dhikr of the destroyer of pleasures (death)." (At-Tirmidhi, an-Nasa'i and Ibn Majah, from Ibn 'Umar)

□ From his companions there are the words of 'Ali ibn Abu Talib ﷺ: "People are asleep; when they die they wake up."

There is no cure except to turn to Allah out of deep need in order to defeat the *nafs*, oppose its whims and direct it to obedience, because the nature of the *nafs* is disobedience. Allah said: *"Truly, the nafs inspires only evil."* (12:53)

THE SHAYKH

303 It is necessary to keep the company of a Shaykh of instruction, who knows the ways, and has completed his own path and freed himself from his passions. The *murid* is the one who submits his self (*nafs*) to the Shaykh, obeying him in everything he instructs, without:

1. Doubts
2. Interpretation and/or
3. Hesitation

The Shaykh serves three functions:

1. Protects from pitfalls of the path (*mahalik at-tariq*).
304 2. Reminds the *murid* of Allah when he sees him.
3. And finally brings the servant to his Master – Allah, the Lord of the Worlds.

The Shaykh must fulfil four conditions:

1. That he knows the individual obligations (*fard 'ayn*) of the Deen according to the fiqh.

2. That he knows the methods of purifying hearts of its diseases.
3. That he has *ma'rifa* (gnosis) – direct experiential knowledge from Allah.
4. That he has permission from his own shaykh (*idhn*) to teach, i.e. that he has a direct chain of transmission to the Imam of the Messengers ﷺ.

The Shaykh of instruction al-Hasan ibn Mas'ud al-Yusi issued a fatwa saying that depending on the state striven for, there are three degrees of need for a Shaykh:

1. For the effort (*mujahada*) to attain *taqwa*, a Shaykh is not a condition, but to have one is better.
2. For the effort to attain *istiqama* ('going straight', i.e. a state of continuous *taqwa* till death), having a Shaykh is also not necessary, but is even more preferable.
3. For the effort to attain *kashf*, by removing (*tajrid*) from the self its vices and its frivolity to be able to reach the truth (*haqiqah*), having a Shaykh is obligatory (*wajib*).[218]

One should follow the *wali* to whom one is led by Allah, and to whose uniqueness He gives access by removing the screen of his humanness. He teaches how to recognise the wrong actions of the *nafs*, in order to avoid them and recognise the favour of Allah towards oneself, so that one flees from other-than-Him to His intimacy.

Mayyara then quotes the well-known saying among the people of tasawwuf: "Whoever has no Shaykh, shaytan is his Shaykh."

Following a teaching Shaykh of Instruction is supported by the following evidence:

1. From Allah's Book:
   - □ *"Follow the way of those who turn back to Me."* (31:14)
   - □ *"You who believe, have taqwa of Allah and be with those who are true."* (9:119)
2. From His Messenger ﷺ that he said:
   - □ "The best people to sit with are those who, when you look at

218 *Shifa' as-Sa'il* by Ibn Khaldun.

them, remind you of Allah, and whose speech increases you in knowledge and whose actions will remind you of the akhira." (Abu Ya'la, from Ibn Abbas; it is *sahih* according to as-Suyuti)

- "The best among you are those who, when they are seen, cause Allah to be remembered." (Al-Hakim at-Tirmidhi, from Anas ibn Malik)

RECKONING (MUHASABA)

305 One must take the self (*nafs*) to account with each breath. Every morning, the *nafs* must be reminded to make use of the new day Allah has granted and every evening, it must reckon all the actions performed. All intended actions must be weighed on the scales of the Shari'a before being performed.

This reckoning relates to two levels of actions:

306
1. The obligatory acts, which serve as base capital and from which any wrong action detracts
2. The supererogatory (*nafila*) deeds (such as sadaqa, fasting and the like) is where profit lies, over and above the base capital, and it is via these supererogatory actions one draws near to Allah

The Messenger ﷺ has explained the above in a *hadith qudsi*:

"Allah says: 'He who is hostile to a friend (*wali*) of Mine, I declare war against. My slave approaches Me with nothing more beloved to Me than that which I have made obligatory for him and My slave keeps drawing nearer to Me with supererogatory (*nafila*) actions until I love him. And when I love him, I am the hearing with which he hears, his sight with which he sees, his hand with which he seizes, and his foot with which he walks. If he asks Me I will surely give to him, and if he seeks refuge in Me, I will surely protect him.'" (*Sahih al-Bukhari*)

DHIKR

307 One must do abundant dhikr with a sincere heart. The word dhikr means simultaneously 'to mention on the tongue' and 'to remember.' Dhikr has three grades:

1. Dhikr with the tongue only
2. Dhikr with the heart only
3. Dhikr with both the tongue and heart together, which is the highest

And the help in doing all of this is from one's Lord.

Doing dhikr is based on the following evidences:

1. From Allah's Book:
   - □ "*Remember Me and I will remember you.*" (2:152)
   - □ "*...Those who remember Allah, standing, sitting and lying down...*" (3:191)
2. From His Messenger ﷺ:
   - □ "There is nothing that the people of the Garden will regret except the moments in which they did not remember Allah, exalted is He." (At-Tabarani, from Mu'adh; it is *hasan* according to as-Suyuti)
   - □ "The likeness of someone who remembers his Lord and someone who does not remember his Lord is the likeness of the living and the dead." (Al-Bukhari)
   - □ "When you pass by the gardens of *Jannah*, graze there." He was asked: "What are the gardens of *Jannah*?" He said: "The circles of dhikr." (At-Tirmidhi, from Anas)
   - □ "Do so much dhikr of Allah until they say you are mad." (Ibn Hanbal)
   - □ "No action done by the child of Adam saves him more from the punishment of Allah, than dhikr of Allah." (Ibn Hanbal, from Mu'adh)

FIGHTING THE SELF

One must strive against one's self (*mujahadat an-nafs*) for the sake of the **308**
Lord of the worlds until it turns from its appetites to Allah's rule, avoiding what is wrong and doing what is right, inwardly and outwardly. This is called the Greater Jihad (*al-jihad al-akbar*).[219]

219 The author of *al-Burda* says: "The nafs is like a child. If it grows up suckling, it loves to suckle, but if it is weaned, it is weaned."

Fighting the self is obligatory because of the following evidence:

1. From Allah's Book: *"As for those who did jihad for Us, We will guide them to our paths."* (29:67) [220]
2. From His Messenger ﷺ:
   - □ Upon returning from battle: "We came back from the lesser jihad to the greater jihad" (Al-Bayhaqi, from Jabir)
   - □ The Messenger ﷺ said: "The *mujahid* is the one who wages *jihad* against his self in [the path of] Allah." (At-Tirmidhi, from Fudala ibn 'Ubaid)

THE PRAISEWORTHY STATES OF THE HEART

The inward obligatory states are called the praiseworthy states of the heart. One must embellish oneself with these stations of certainty (*maqamat al-yaqin*):

309 1. Fear (*khawf*). Its evidence is:
   - □ From the Book of Allah:
     - *"For those who fear the station of their Lord, there are two gardens."* (55:46)
     - *"Only those of His slaves with knowledge have fear of Allah."* (35:28)
   - □ From His Messenger ﷺ: "By Allah, I am the one with most fear and *taqwa* of Allah among you." (Al-Bukhari, from Anas)

2. Hope (*raja'*), but letting fear dominate (except when ill, then hope should dominate)

   Its evidence is:

   - □ From the Book of Allah: *"Those who believed, and those who performed hijra and made jihad in the way of Allah, they hope for Allah's mercy."* (2:218)
   - □ From His Messenger ﷺ: the hadith *qudsi*: "Son of Adam! If you worship Me, hope for Me and you do not associate anything with Me, I will forgive you regardless of what you

---

220 This surah, al-'Ankabut, is a Makkan surah and military jihad was only legislated in the Madinan period. This is evidence that the jihad mentioned in this verse is the jihad of the self, according to Ibn Juzayy in his tafsir.

have done. If you were to face Me with wrong actions the size of the heaven and the earth, I will face you with forgiveness the size of the heaven and the earth. I will forgive you, and I will not care." (At-Tabarani, from Abu Darda' and declared *hasan* by as-Suyuti)

3. Gratitude (*shukr*) for the gifts granted

   Its evidence is:

   - From the Book of Allah: "*We will recompense the grateful.*" (3:145)
   - From the Messenger's ﷺ reply when asked why he stood in salat till his feet hurt though Allah had forgiven him past and future wrongs: "Should I not be a grateful slave?" (Agreed upon, from al-Mughira ibn Shu'ba)

4. Steadfastness (*sabr*) in afflictions

   Its evidence is:

   - From the Book of Allah: "*Allah loves the steadfast.*" (3:146)
   - From His Messenger ﷺ, "No-one has received a better and greater gift than steadfastness." (Agreed upon, from Abu Sa'id)

5. Turning to Allah (*tawba*)[221]
6. Doing without in the dunya (*zuhd*) and preferring the akhira, so that one only takes from the dunya what is necessary

   Its evidence is:

   - From the Book of Allah: "*Whatever you have will end, but what Allah has is lasting.*" (16:96)
   - From His Messenger ﷺ: "Do without the dunya and Allah will love you. Do without what is in people's hands, and people will love you." (Ibn Maja, at-Tabarani, al-Hakim, al-Bayhaqi, from Sahl ibn Sa'd and declared *sahih* by as-Suyuti)

7. Reliance (*tawakkul*) in all affairs

   Its evidence is:

   - From the Book of Allah:

221 See section on *tawba* for an explanation and evidences.

- *"Let the those who have Iman rely on Allah!"* (3:122)
- *"Those to whom people said, 'The people have gathered against you, so fear them.' But it only increased their Iman, and they said: 'Allah is sufficient for us, and He is the Best of Guardians!'"* (3:173)

8. Satisfaction (*rida*) with what Allah has decreed, the sweet and bitter of it

Its evidence is:

□ From Allah's Book: *"Allah is satisfied with them and they are satisfied with Him."* (5:119)

□ From His Messenger ﷺ: "Allah will say, 'Oh inhabitants of *Jannah*!' and they will respond 'Here we are Lord, good is in Your hands!' and He will ask, 'Are you satisfied?' and they will reply, 'How can we not be satisfied? You have given to us what you have not given to anyone else in Your creation!' And He will say, 'Should I give you what is better than this?' And they will say, What is better than this?' And He will say: "That My satisfaction has been granted to you and I will never again be dissatisfied with you." (Agreed upon, from Abu Sa'id al-Khudri)

9. Love (*hubb/mahabbah*) of Allah and love of His Messenger ﷺ, which is the core of love of Allah (and includes loving the 'ulama and the people of *taqwa*) because the beloved of the beloved is beloved. It brings satisfaction with everything that is from Him, because love brings along satisfaction with the actions of the beloved.

Its evidence is:

□ From Allah's Book:

- *"...He will bring about a people whom He loves, and they love Him."* (5:54)
- His order to his Messenger ﷺ: *"Say: 'If you love Allah, follow me and Allah will love you and forgive you your wrong actions.'"*(3:31)
- *"Say: 'If your fathers or your sons or your brothers or your wives or your tribe, or any wealth you have acquired, or any business you fear may slump, or any dwelling-places which please you, are dearer to you than Allah and His Messenger*

and doing jihad in His Way, then wait until Allah brings about His command. Allah does not guide people who are transgressors.'" (9:24)

- ☐ From Allah's Messenger ﷺ:
  - "There are three things which cause anyone who takes refuge in them to experience the sweetness of belief – that Allah and His Messenger are more beloved to him than anything else; that he loves a man only for Allah; and that he dislikes the thought of reverting to disbelief as much as he would dislike being cast into the Fire." (Agreed upon, from Anas)
  - "'Umar ibn al-Khattab told the Prophet ﷺ, "I love you more than anything except my soul which is between my two sides." The Prophet ﷺ replied, "None of you will believe until I am dearer to him than his own soul." 'Umar said, "By the One Who sent down the Book on you, I love you more than my soul which is between my two sides." The Prophet ﷺ said, "'Umar, now you have it!" (Al-Bukhari, from 'Abdullah ibn Hisham)

10. Sincerity (*sidq*). The slave must do all his acts of obedience for Allah's **310**
sake alone, Who is watching him inwardly and outwardly, not for prestige, showing off, etc.

    Its evidence is:

    - ☐ From Allah's Book: "*They were only ordered to worship Allah, making their* Deen *sincerely His.*" (98:5)
    - ☐ From Allah's Messenger ﷺ, "Allah does not look at your bodies or your forms, but at your hearts." (Muslim, from Abu Hurayra)

We have already spoken about satisfaction with what Allah has decreed.

THE BENEFITS OF TASAWWUF

All the Islamic sciences have their benefits. Ibn 'Ashir mentions four benefits of travelling the path of tasawwuf:

1. Gnosis (*ma'rifa*) is overwhelming experiential knowledge and a **311**

gift directly bestowed by the Lord of the Worlds, whereas *'ilm* is information acquired by study and learning. Gnosis is a result of action, as the Messenger ﷺ said: "Whoever acts on what he knows, Allah will grant him a knowledge he did not know."[222] It has two levels:

- □ The knowledge possessed by all *muminun* which is the knowledge of Allah and His Messenger ﷺ, i.e. the Shahada, which is in itself an immense opening
- □ The knowledge possessed by the prophets and *'awliya*, and this is a boundless ocean. Shaykh Ibn 'Ata'illah said: "Whoever knows the real will contemplate Him in everything. Whoever is annihilated in Him is absent from everything, and whoever loves Him prefers Him over everything."[223]

2. Freedom (*hurriyya*) from everything except Allah. Shaykh Ibn 'Ata'Illah said: "There is nothing that you love, except you are a slave of it, and He does not want you to be a slave of anyone but Him."[224] And "You are free of what you despair of and you are a slave of what you yearn for."[225]

312 3. Allah's love (*hubb*). We have already provided evidences for this station. It is described in *al-Habl al-Matin* quoting from the *Ihya' 'Ulum ad-Din* that love of Allah for His slave is the drawing near of the slave to Him by:
   - □ Removing everything that distracts the slave from Him
   - □ Removing from him all transgressions
   - □ Cleaning him from all worldly turbidity
   - □ Removing his veil until he witnesses Him as if he sees Him with his heart

4. Allah choosing His servant to enter His Pure Presence (*Hadrat al-Quds*). This is what is referred to in the famous hadith of Jibril عليه السلام when he asks the Prophet ﷺ "what is Ihsan" and the Prophet ﷺ answers "To worship Allah as if you see Him." Entering into His Pure Presence is the station of *mushahada* or contemplation of His Pure Unity (*tawhid*) as expressed in the hadith.

---

222 Abu Nu'aym, from Anas.

223 *Al-Hikam*, aphorism 163.

224 *Al-Hikam*, aphorism 210.

225 *Al-Hikam*, aphorism 62.

Know – may Allah have mercy on all of us – that *tawhid* is of two types:

- The general *tawhid* that all Muslims have, which is avoiding open *shirk* (associating or ascribing partners to Allah)
- The *tawhid* of the élite, which is avoiding hidden *shirk*

Hidden *shirk* is to consider the self (*nafs*) the centre and basis of everything, whereas true *tawhid* is to realise that human existence is metaphorical (*wujud majazi*)[226] and only Allah has real existence (*wujud haqiqi*).[227]

So Ihsan is to know that Allah is the centre and the measure of everything. What is relevant is what He wants, not what the self wants.

Tasawwuf is the science of dismantling the self until nothing remains but witnessing the Endless Ocean of Unity: that Allah truly is One – with no co-sharer in his Essence, Attributes and Actions.[228]

"Everyone on it disappears, and there will remain the Face of your Lord, owner of Majesty and Generosity." (55:26-27).

---

226 Meaning that human existence has no actuality in itself and is completely dependent on the Existence of Allah.

227 Meaning that Allah's Existence is independent of all else and is absolute.

228 As the teacher of my teacher, Shaykh Muhammad ibn al-Habib al-Hasani, said, "The science of *kalam* ends where the science of tasawwuf starts, at the shore of the sea".

End

313 THE AUTHOR SAYS that the teachings contained in this poem do not encompass everything necessary that a person has to learn about the Deen. What a person has to learn about the Deen is more than what is contained in these verses, yet what has been mentioned will be enough for the one who devotes himself to learn it.[229]

314 The author has managed to gather all this knowledge in 314 verses, which coincides with the number of Messengers.

315 The author has named his poem *al-Murshid al-Mu'in 'ala ad-Daruri min 'Ulum ad-Deen* – "The Helpful Guide to the Necessary Sciences of the Deen."[230]

316 Then, he asks Allah to continuously benefit people by this poem by the noble rank of the Master of Mankind, Sayyiduna Muhammad ﷺ.

317 Then he seals the poem in the same way that he started it, by praising Allah, exalted is He, and by asking for blessings and peace upon the most noble and generous guide, Sayyiduna Muhammad ﷺ.

THIS WORK WAS completed on the night of the 14th of Rabi' al-Awwal 1431, corresponding to 28th of February 2010 by the slave of Allah, much in need of His mercy, 'Ali ibn Abdil-Haqq al-'Iraqi al-Husaini and his student the slave of Allah, Yakub Werdelin. May Allah be pleased with them and forgive them their wrong actions and mistakes.

We ask Allah to give benefit by it as He has given benefit by its source, and to accept it from us and to put it in the scale of accepted actions. Amin.

---

229 If a person learns and puts what is in the poem into practice, Allah will teach him what he did not know, and this is something verified.

230 The title of the poem truly expresses the nature of the text.

اَلْمُرْشِدُ الْمُعِينُ

عَلَى الضَّرُورِي مِن عُلُومِ الدِّينِ

1. يَقُولُ عَبْدُ الْوَاحِدِ بْنُ عَاشِرِ مُبْتَدِئاً بِاسْمِ الْإِلَهِ الْقَادِرِ

2. اَلْحَمْدُ لِلَّهِ الَّذِي عَلَّمَنَا مِنَ الْعُلُومِ مَا بِهِ كَلَّفَنَا

3. صَلَّى وَسَلَّمَ عَلَى مُحَمَّدِ وَآلِهِ وَصَحْبِهِ وَالْمُقْتَدِي

4. وَبَعْدُ فَالْعَوْنُ مِنَ اللَّهِ الْمَجِيدْ فِي نَظْمِ أَبْيَاتٍ لِلأُمِّيِّ تُفِيدْ

5. فِي عَقْدِ الْأَشْعَرِي وفِقْهِ مَالِكْ وَفِي طَرِيقَةِ الْجُنَيْدِ السَّالِكْ

مُقَدِّمَةٌ لِكِتَابِ الْاعْتِقَادِ مُعِينَةٌ لِقَارِئِهَا عَلَى الْمُرَادِ

6. وَحُكْمُنَا الْعَقْلِي قَضِيَّةٌ بِلاَ وَقْفٍ عَلَى عَادَةٍ أَوْ وَضْعٍ جَلاَ

7. أَقْسَامُ مُقْتَضَاهُ بِالْحَصْرِ تُمَازْ وهْيَ الْوُجُوبُ الِاسْتِحَالَةُ الْجَوَازْ

8. فَوَاجِبٌ لاَ يَقْبَلُ النَّفْيَ بِحَالْ ومَا أَبَى الثُّبُوتَ عَقْلاً الْمُحَالْ

9. وجَائِزاً مَاقَبِلَ الأَمْرَيْنِ سِمْ لِلضَّرُورِي والنَّظَرِي كُلٌّ قُسِمْ

10. أَوَّلُ وَاجِبٍ عَلَى مَنْ كُلِّفَا   مُمَكَّناً مِنْ نَظَرٍ أنْ يَعْرِفَا

11. اللَّهَ وَالرُّسُلَ بِالصِّفَاتِ   مِمَّا عَلَيْهِ نَصَبَ الآيَاتِ

12. وَكُلُّ تَكْلِيفٍ بِشَرْطِ الْعَقْلِ   مَعَ الْبُلُوغِ بِدَمٍ أَوْ حَمْلِ

13. أَوْ بِمَنِيٍّ أَو بِإِنْبَاتِ الشَّعَرِ   أَوْ بِثَمَانِ عَشْرَةٍ حَوْلاً ظَهَرْ

كِتَابُ أُمِّ الْقَوَاعِدِ وَمَا انْطَوَتْ عَلَيْهِ مِنَ الْعَقَائِدِ

14. يَجِبُ لِلَّهِ الْوُجُودُ وَالْقِدَمْ   كَذَا الْبَقَاءُ وَالْغِنَى الْمُطْلَقُ عَمْ

15. وَخُلْفُهُ لِخَلْقِهِ بِلاَ مِثَالْ   وَوَحْدَةُ الذَّاتِ وَوَصْفٍ وَالْفِعَالْ

16. وَقُدْرَةٌ إِرَادَةٌ عِلْمٌ حَيَاةْ   سَمْعٌ كَلاَمٌ بَصَرٌ ذِي وَاجِبَاتْ

17. وَيَسْتَحِيلُ ضِدُّ هَذِهِ الصِّفَاتْ   الْعَدَمُ الْحُدُوثُ ذَا لِلْحَادِثَاتِ

18. كَذَا الْفَنَا وَالاِفْتِقَارُ عُدَّهْ   وَأَنْ يُمَاثَلُ وَنَفْيُ الْوَحْدَةْ

19. عَجْزٌ كَرَاهَةٌ وَجَهْلٌ وَمَمَاتْ   وَصَمَمٌ وَبَكَمٌ عَمًى صُمَاتْ

20. يَجُوزُ فِي حَقِّهِ فِعْلُ الْمُمْكِنَاتْ   بِأَسْرِهَا وَتَرْكُهَا فِي الْعَدَمَاتْ

21. وُجُودُهُ لَهُ دَلِيلٌ قَاطِعْ   حَاجَةُ كُلِّ مُحْدَثٍ لِلصَّانِعْ

22. لَوْ حَدَثَتْ بِنَفْسِهَا الأَكْوَانُ   لاَجْتَمَعَ التَّسَاوِي وَالرُّجْحَانُ

23. وَذَا مُحَالٌ وَحُدُوثُ الْعَالَمِ   مِنْ حَدَثِ الأَعْرَاضِ مَعْ تَلاَزُمِ

24. لَوْ لَمْ يَكُ الْقِدَمُ وَصْفَهُ لَزِمْ   حُدُوثُهُ دَوْرٌ تَسَلْسُلٌ حُتِمْ

25. لَوْ أَمْكَنَ الْفَنَاءُ لاَنْتَفَى الْقِدَمْ   لَوْ مَاثَلَ الْخَلْقَ حُدُوثُهُ انْحَتَمْ

26. لَوْ لَمْ يَجِبْ وَصْفُ الْغِنَى لَهُ افْتَقَرْ   لَوْ لَمْ يَكُنْ بِوَاحِدٍ لَمَا قَدَرْ

27. لَوْ لَمْ يَكُنْ حَيّاً مُرِيداً عَالِمَا   وَقَادِراً لَمَا رَأَيْتَ عَالَمَا

28. وَالتَّالِ فِي السِّتِّ الْقَضَايَا بَاطِلْ   قَطْعاً مُقَدَّمٌ إِذاً مُمَاثِلْ

29. وَالسَّمْعُ وَالْبَصَرُ وَالْكَلاَمْ   بِالنَّقْلِ مَعْ كَمَالِهِ تُرَامْ

30. لَوْ اسْتَحَالَ مُمْكِنٌ أَوْ وَجَبَا   قَلْبُ الْحَقَائِقِ لُزُوماً أَوْجَبَا

31. يَجِبُ لِلرُّسْلِ الْكِرَامِ الصِّدْقُ   أَمَانَةٌ تَبْلِيغُهُمْ يَحِقُّ

32. مُحَالٌ الْكَذِبُ وَالْمَنْهِيُّ   كَعَدَمِ التَّبْلِيغِ يَا ذَكِيُّ

33. يَجُوزُ فِي حَقِّهِمْ كُلُّ عَرَضْ   لَيْسَ مُؤَدِّياً لِنَقْصٍ كَالْمَرَضْ

34. لَوْ لَمْ يَكُونُوا صَادِقِينَ لَلَزِمْ   أَنْ يَكْذِبَ الإِلَهُ فِي تَصْدِيقِهِمْ

35. إِذْ مُعْجِزَاتُهُمْ كَقَوْلِهِ وَبَرّ   صَدَقَ هَذَا الْعَبْدُ فِي كُلِّ خَبَرْ

36. لَوِ انْتَفَى التَّبْلِيغُ أَوْ خَانُوا حُتِمْ   أَنْ يُقْلَبَ الْمَنْهِيُّ طَاعَةً لَهُمْ

37. جَوَازُ الأَعْرَاضِ عَلَيْهِمْ حُجَّتُهْ   وُقُوعُهَا بِهِمْ تَسَلٍّ حِكْمَتُهْ

38. وَقَوْلُ لاَ إِلَهَ إِلاَّ اللَّهُ   مُحَمَّدٌ أَرْسَلَهُ الْإِلَهُ

39. يَجْمَعُ كُلَّ هَذِهِ الْمَعَانِي   كَانَتْ لِذَا عَلاَمَةَ الْإِيمَانِ

40. وَهِيَ أَفْضَلُ وُجُوهِ الذِّكْرِ   فَاشْغَلْ بِهَا الْعُمْرَ تَفُزْ بِالذُّخْرِ

فَصْلٌ فِي قَوَاعِدِ الْإِسْلَامِ

41. **فَصْلٌ** وَطَاعَةُ الْجَوَارِحِ الْجَمِيعْ   قَوْلاً وَفِعْلاً هُوَ الإِسْلاَمُ الرَّفِيعْ

42. قَوَاعِدُ الإِسْلاَمِ خَمْسٌ وَاجِبَاتْ   وَهْيَ الشَّهَادَتَانِ شَرْطُ البَاقِيَاتْ

43. ثُمَّ الصَّلاةُ والزَّكَاةُ فِي الْقِطَاعْ   والصَّوْمُ والحَجُّ عَلَى مَنِ اسْتَطَاعْ

44. الْإِيمَانُ جَزْمٌ بِالإِلَهِ وَالْكُتُبْ   وَالرُّسْلِ وَالأَمْلاَكِ مَعْ بَعْثٍ قَرُبْ

45. وقَدَرٌ كَذَا الصِّرَاطُ فَالْمِيزَانْ   حَوْضُ النَّبِيِّ جَنَّةٌ وَنِيرَانْ

46. وَأَمَّا الْإِحْسَانُ فَقَالَ مَنْ دَرَاهْ   أَنْ تَعْبُدَ اللَّهَ كَأَنَّكَ تَرَاهْ

47. إِنْ لَمْ تَكُنْ تَرَاهُ إِنَّهُ يَرَاكْ   وَالدِّينُ ذِي الثَّلاَثِ خُذْ أَقْوَى عُرَاكْ

مُقَدِّمَةٌ مِنَ الأُصُولِ مُعِينَةٌ فِي فُرُوعِهَا عَلَى الْوُصُولِ

48. اَلْحُكْمُ فِي الشَّرْعِ خِطَابُ رَبِّنَا   الْمُقْتَضِي فِعْلَ الْمُكَلَّفِ افْطُنَا

49. بِطَلَبٍ أَوْإِذْنٍ أَوْبِوَضْعِ   لِسَبَبٍ أَوْشَرْطٍ أَوْذِي مَنْعِ

50. أَقْسَامُ حُكْمِ الشَرْعِ خَمْسَةٌ تُرَامْ   فَرْضٌ وَنَدْبٌ وَكَرَاهَةٌ حَرَامْ

51. ثُمَّ إِبَاحَةٌ فَمَأْمُورٌ جُزِمِ   فَرْضٌ وَدُونَ الْجَزْمِ مَنْدُوبٌ وُسِمَ

52. ذُو النَّهْيِ مَكْرُوهٌ وَمَعْ حَتْمٍ حَرَامْ   مَأْذُونُ وَجْهَيْهِ مُبَاحٌ ذَا تَمَامْ

53. وَالْفَرْضُ قِسْمَانِ كِفَايَةٌ وَعَيْنْ   وَيَشْمَلُ المَنْدُبُ سُنَّةً بِذَيْنْ

كِتَابُ الطَّهَارَةِ

54. **فَصْلٌ** وَتَحْصُلُ الطَّهَارَةُ بِمَا   مِنَ التَّغَيُّرِ بِشَيْءٍ سَلِمَا

55. إِذَا تَغَيَّرَ بِنَجْسٍ طُرِحَا   أَوْ طَاهِرٍ لِعَادَةٍ قَدْ صَلُحَا

56. إِلاَّ إِذَا لاَزَمَهُ فِي الْغَالِبِ   كَمُغْرَةٍ فَمُطْلَقٌ كَالذَّائِبِ

فَرَائِضُ الْوُضُوءِ

57. **فَصْلٌ** فَرَائِضُ الْوُضُوءِ سَبْعٌ وَهِيَ   دَلْكٌ وَفَوْرٌ نِيَّةٌ فِي بَدْئِهِ

58. وَلْيَنْوِ رَفْعَ حَدَثٍ أَوْ مُفْتَرَضْ   أَوْ اسْتِبَاحَةً لِمَمْنُوعٍ عَرَضْ

59. وغَسْلُ وَجْهٍ غَسْلُهُ الْيَدَيْنِ   وَمَسْحُ رَأْسٍ غَسْلُهُ الرِّجْلَيْنِ

60. وَالْفَرْضُ عَمَّ مَجْمَعَ الأُذْنَيْنِ   وَالْمَرْفِقَيْنِ عَمَّ وَالْكَعْبَيْنِ

61. خَلِّلْ أَصَابِعَ الْيَدَيْنِ وَشَعَرْ   وَجْهٍ إِذَا مَا تَحْتَهُ الْجِلْدُ ظَهَرْ

سُنَنُ الْوُضُوءِ

62. سُنَنُهُ السَّبْعُ ابْتِدَا غَسْلُ الْيَدَيْنِ   وَرَدُّ مَسْحِ الرَّأْسِ مَسْحُ الأُذُنَيْنِ

63. مَضْمَضَةٌ اسْتِنْشَاقٌ اسْتِنْثَارُ   تَرْتِيبُ فَرْضِهِ وَذَا الْمُخْتَارُ

64. وَأَحَدَ عَشْرَ الْفَضَائِلُ أَتَتْ   تَسْمِيَةٌ وبُقْعَةٌ قَدْ طَهُرَتْ

65. تَقْلِيلُ مَاءٍ وَتَيَامُنُ الإِنَا   وَالشَّفْعُ وَالتَّثْلِيثُ فِي مَغْسُولِنَا

66. بَدْءُ الْمَيَامِنِ سِوَاكٌ وَنُدِبْ   تَرْتِيبُ مَسْنُونِهِ أَوْ مَعَ مَا يَجِبْ

67. وَبَدْءُ مَسْحِ الرَّأْسِ مِنْ مُقَدَّمِهْ   تَخْلِيلُهُ أَصَابِعاً بِقَدَمِهْ

68. وَكُرِهَ الزَّيْدُ عَلَى الْفَرْضِ لَدَى   مَسْحٍ وَفِي الْغَسْلِ عَلَى مَاحُدِّدَا

69. وَعَاجِزُ الْفَوْرِ بَنَى مَالَم يَطُلْ   بِيُبْسِ الأَعْضَا فِي زَمَانٍ مُعْتَدِلْ

70. ذَاكِرُ فَرْضِهِ بِطُولٍ يَفْعَلُهْ   فَقَطْ وَفِي الْقُرْبِ الْمُوَالِي يُكْمِلُهْ

71. إِنْ كانَ صَلَّى بَطَلَتْ وَمَنْ ذَكَرْ سُنَّتَهُ يَفْعَلُهَا لِمَا حَضَرْ

نَوَاقِضُ الْوُضُوءِ

72. نَوَاقِضُ الْوُضُوءِ سِتَّةَ عَشَرْ بَوْلٌ وَرِيحٌ سَلَسٌ إِذَا نَدَرْ

73. وَغَائِطٌ نَوْمٌ ثَقِيلٌ مَذْيُ سُكْرٌ وَإِغْماءٌ جُنُونٌ وَدْيُ

74. لَمْسٌ وَقُبْلَةٌ وَذَا إِنْ وُجِدَتْ لَذَّةُ عادَةٍ كَذَا إِنْ قُصِدَتْ

75. إِلْطَافُ مَرْأَةٍ كَذا مَسُّ الذَّكَرْ وَالشَّكُّ فِي الْحَدَثِ كُفْرُ مَنْ كَفَرْ

76. وَيَجِبُ اسْتِبْرَاءُ الأَخْبَثَيْنِ مَعْ سَلْتِ وَنَتْرِ ذَكَرٍ وَالشَّدَّ دَعْ

77. وَجازَ الاِسْتِجْمَارُ مِنْ بَوْلِ ذَكَرْ كَغَائِطٍ لاَ مَا كَثِيراً انْتَشَرْ

فَرَائِضُ الْغُسْلِ

78. **فَصْلٌ** فُرُوضُ الغُسلِ قَصْدٌ يُحْتَضَرْ فَوْرٌ عُمُومُ الدَّلْكِ تَخْلِيلُ الشَّعَرْ

79. فَتابِعِ الْخَفِيَّ مِثْلَ الرُّكْبَتَيْنِ الإِبْطِ وَالرُّفْغِ وَبَيْنَ الأَلْيَتَيْنِ

80. وَصِلْ لِمَا عَسُرَ بِالْمِنْدِيلِ وَنَحْوِهِ كَالْحَبْلِ وَالتَّوْكِيلِ

81. سُنَنُهُ مَضْمَضَةٌ غَسْلُ الْيَدَيْنِ بَدْءًا وَالاِسْتِنْشَاقُ ثُقْبُ الأُذُنَيْنِ

82. مَنْدُوبُهُ الْبَدْءُ بِغَسْلِهِ الأَذَى تَسْمِيَةٌ تَثْلِيثُ رَأْسِهِ كَذَا

83. تَقْدِيمُ أَعْضَاءِ الْوُضُوءِ قِلَّةُ مَا — بَدْءٌ بِأَعْلَى وَيَمِينٍ خُذْهُمَا

84. تَبْدَأُ فِي الْغُسْلِ بِفَرْجٍ ثُمَّ كُفْ — عَنْ مَسِّهِ بِبَطْنٍ أَو جَنَبِ الأَكُفْ

85. أَو إِصْبُعٍ ثُمَّ إِذَا مَسَسْتَهْ — أَعِدْ مِنَ الْوُضُوءِ مَا فَعَلْتَهْ

86. مُوجِبُهُ حَيْضٌ نِفَاسٌ إِنْزَالْ — مَغِيبُ كَمْرَةٍ بِفَرْجٍ إِسْجَالْ

87. وَالأَوَّلَانِ مَنَعَا الْوَطْءَ إِلى — غُسْلٍ وَالآخَرَانِ قُرْآنًا حَلاَ

88. وَالكُلَّ مَسْجِدًا وَسَهْوُ الاِغْتِسالْ — مِثْلَ وُضُوئِكَ وَلَمْ تُعِدْ مُوَالْ

التَّيَمُّمُ

89. **فَصْلٌ** لِخَوفِ ضُرٍّ أَوْ عَدَمِ مَا — عَوِّضْ مِنَ الطَّهَارَةِ التَّيَمُّمَا

90. وَصَلِّ فَرْضاً وَاحِداً وَإِنْ تَصِلْ — جَنَازَةً وَسُنَّةً بِهِ يَحِلْ

91. وَجَازَ لِلنَّفْلِ ابْتِداً وَيَسْتَبِيحْ — الْفَرْضَ لاَ الْجُمْعَةَ حَاضِرٌ صَحِيحْ

فَرَائِضُ التَّيَمُّمُ

92. فُرُوضُهُ مَسْحُكَ وَجْهًا وَالْيَدَيْنِ — لِلْكُوعِ وَالنِّيَّةُ أُولَى الضَّرْبَتَيْنِ

93. ثُمَّ الْمُوَالاَةُ صَعِيدٌ طَهُرَا — وَوَصْلُهَا بِهِ وَوَقْتٌ حَضَرَا

94. آخِرُهُ لِلرَّاجِي آيِسٌ فَقَطْ — أَوَّلُهُ وَالْمُتَرَدِّدُ الْوَسَطْ

سُنَنُ التَّيَمُّمُ

95. سُنَنُهُ مَسْحُهُمَا لِلْمِرْفَقِ وضَرْبَةُ الْيَدَيْنِ تَرْتِيبٌ بَقِي

96. مَنْدُوبُهُ تَسْمِيَةٌ وَصْفٌ حَمِيدْ نَاقِضُهُ مِثْلُ الْوُضُوءِ وَيَزِيدْ

97. وُجُودُ مَاءٍ قَبْلَ أَنْ صَلَّى وَإِنْ بَعْدُ يَجِدْ يُعِدْ بِوَقْتٍ إِنْ يَكُنْ

98. كَخَائِفِ اللِّصِّ وَرَاجٍ قَدَّمَا وَزَمِنٍ مُنَاوِلاً قَدْ عَدِمَا

كِتَابُ الصَّلَاةِ

99. **فَرَائِضُ** الصَّلَاةِ سِتَّ عَشَرَهْ شُرُوطُهَا أَرْبَعَةٌ مُفْتَقِرَهْ

100. تَكْبِيرَةُ الاِحْرَامِ وَالْقِيَامُ لَهَا وَنِيَّةٌ بِهَا تُرَامْ

101. فَاتِحَةٌ مَعَ الْقِيَامِ وَالرُّكُوعْ وَالرَّفْعُ مِنْهُ وَالسُّجُودُ بِالْخُضُوعْ

102. وَالرَّفْعُ مِنْهُ وَالسَّلاَمُ وَالْجُلُوسْ لَهُ وَتَرْتِيبُ أَدَاءٍ فِي الأُسُوسْ

103. وَالاِعْتِدَالُ مُطْمَئِنًّا بِالْتِزَامْ تَابَعَ مَأْمُومٌ بِإِحْرَامٍ سَلاَمْ

104. نِيَّتُهُ اقْتِدَاءٍ كَذَا الإِمَامُ فِي خَوْفٍ وَجَمْعٍ جُمُعَةٍ مُسْتَخْلَفٍ

105. شَرْطُهَا الاِسْتِقْبَالُ طُهْرُ الْخَبَثِ وَ سَتْرُ عَوْرَةٍ وَطُهْرُ الْحَدَثِ

106. بِالذِّكْرِ وَالْقُدْرَةِ فِي غَيْرِ الأَخِيرْ   تَفْرِيعُ نَاسِيهَا وَعَاجِزٌ كَثِيرْ

107. نَدْبًا يُعِيدَانِ بِوَقْتٍ كَالْخَطَا   فِي قِبْلَةٍ لاَ عَجْزِهَا أَوِ الْغِطَا

108. وَمَا عَدَا وَجْهَ وَكَفَّ الْحُرَّةِ   يَجِبُ سَتْرُهُ كَمَا فِي الْعَوْرَةِ

109. لَكِنْ لَدَى كَشْفٍ لِصَدْرٍ أَوْ شَعَرْ   أَوْ طَرَفٍ تُعِيدُ فِي الْوَقْتِ الْمُقَرْ

110. شَرْطُ وُجُوبِهَا النَّقَا مِنَ الدَّمِ   بِقَصَّةٍ أَوِ الْجُفُوفِ فَاعْلَمِ

111. فَلاَ قَضَا أَيَّامَهُ ثُمَّ دُخُولْ   وَقْتٍ فَأَدِّهَا بِهِ حَتْمًا أَقُولْ

سُنَنُ الصَّلاَةِ

112. سُنَنُهَا السُّورَةُ بَعْدَ الْوَاقِيَهْ   مَعَ الْقِيَامِ أَوَّلاً وَالثَّانِيَهْ

113. جَهْرٌ وَسِرٌّ بِمَحَلٍّ لَهُمَا   تَكْبِيرُهُ إِلاَّ الَّذِي تَقَدَّمَا

114. كُلُّ تَشَهُّدٍ جُلُوسٌ أَوَّلُ   وَالثَّانِي لاَ مَا لِلسَّلاَمِ يَحْصُلُ

115. وَسَمِعَ اللَّهُ لِمَنْ حَمِدَهْ   فِي الرَّفْعِ مِنْ رُكُوعِهِ أَوْرَدَهْ

116. اَلْفَذُّ وَالإِمَامُ هَذَا أُكِّدَا   وَالْبَاقِي كَالْمَنْدُوبِ فِي الْحُكْمِ بَدَا

117. إِقَامَةٌ سُجُودُهُ عَلَى الْيَدَيْنِ   وَطَرَفُ الرِّجْلَيْنِ مِثْلُ الرُّكْبَتَيْنِ

118. إِنْصَاتُ مُقْتَدٍ بِجَهْرٍ ثُمَّ رَدْ   عَلَى الإِمَامِ وَالْيَسَارِ وَأَحَدْ

119. بِهِ وَزَائِدُ سُكُونٍ لِلْحُضُورْ سُتْرَةُ غَيْرِ مُقْتَدٍ خَافَ الْمُرُورْ

120. جَهْرُ السَّلاَمِ كَلِمُ التَّشَهُّدِ وَأَنْ يُصَلِّى عَلَى مُحَمَّدِ

121. سُنَّ الأَذَانُ لِجَمَاعَةٍ أَتَتْ فَرْضاً بِوَقْتِهِ وَغَيْراً طَلَبَتْ

122. وَقَصْرُ مَنْ سَافَرَ أَرْبَعَ بُرُدْ ظُهْراً عِشَا عَصْراً إِلَى حِينِ يَعُدْ

123. مِمَّا وَرَا السُّكْنَى إِلَيْهِ إِنْ قَدِمْ مُقِيمُ أَرْبَعَةِ أَيَّامٍ يُتِمْ

مَنْدُوبَاتُ الصَّلاَةِ

124. مَنْدُوبُهَا تَيَامُنٌ مَعَ السَّلاَمْ تَأْمِينُ مَنْ صَلَّى عَدَا جَهْرِ الإِمَامْ

125. وَقَوْلُ رَبَّنَا لَكَ الْحَمْدُ عَدَا مَنْ أَمَّ وَالْقُنُوتُ فِي الصُّبْحِ بَدَا

126. رِدًا وَتَسْبِيحُ السُّجُودِ وَالرُّكُوعْ سَدْلُ يَدٍ تَكْبِيرُهُ مَعَ الشُّرُوعْ

127. وَبَعْدَ أَنْ يَقُومَ مِنْ وُسْطَاهُ وَعَقْدُهُ الثَّلاَثَ مِنْ يُمْنَاهُ

128. لَدَى التَّشَهُّدِ وَبَسْطُ مَاخَلاَهْ تَحْرِيكُ سَبَّابَتِهَا حِينَ تَلاَهْ

129. وَالْبَطْنُ مِنْ فَخِذٍ رِجَالٌ يُبْعِدُونْ وَمِرْفَقًا مِنْ رُكْبَةٍ إِذْ يَسْجُدُونْ

130. وَصِفَةُ الْجُلُوسِ تَمْكِينُ الْيَدِ مِنْ رُكْبَتَيْهِ فِي الرُّكُوعِ وَزِدْ

131. نَصْبُهُمَا قِرَاءَةُ الْمَأْمُومِ فِي سِرِّيَّةٍ وَضْعُ الْيَدَيْنِ فَاقْتَفِي

132. لَدى السُّجُودِ حَذْوَ أُذْنٍ وَكَذَا رَفْعُ الْيَدَيْنِ عِنْدَ الْإِحْرَامِ خُذَا

133. تَطْوِيلُهُ صُبْحاً وَظُهْراً سُورَتَيْنِ تَوَسُّطُ الْعِشَا وَقَصْرُ الْبَاقِيَيْنِ

134. كَالسُّورَةِ الأُخْرَى كَذَا الْوُسْطَى اسْتُحِبْ سَبْقُ يَدٍ وَضْعاً وَفِي الرَّفْعِ الرُّكَبْ

135. وَكَرِهُوا بَسْمَلَةً تَعَوُّذاً فِي الْفَرْضِ وَالسُّجُودَ فِي الثَّوْبِ كَذَا

136. كَوْرُ عِمَامَةٍ وَبَعْضُ كُمِّهِ وَحَمْلُ شَيْءٍ فِيهِ أَوْ فِي فَمِهِ

137. قِرَاءَةٌ لَدَى السُّجُودِ وَالرُّكُوعْ تَفَكُّرُ الْقَلْبِ بِمَا نَافَى الْخُشُوعْ

138. وَعَبَثٌ وَالِالْتِفَاتُ وَالدُّعَا أَثْنَاءَ قِرَاءَةٍ كَذَا إِنْ رَكَعَا

139. تَشْبِيكٌ أَوْ فَرْقَعَةُ الْأَصَابِعْ تَخَصُّرُ تَغْمِيضُ عَيْنٍ تَابِعْ

140. فَصْلٌ وَخَمْسُ صَلَوَاتٍ فَرْضُ عَيْنْ وَ هْيَ كِفَايَةٌ لِمَيْتٍ دُونَ مَيْنْ

141. فُرُوضُهَا التَّكْبِيرُ أَرْبَعاً دُعَا وَنِيَّةٌ سَلاَمُ سِرٍّ تَبِعَا

142. وَكَالصَّلاَةِ الْغُسْلُ دَفْنُ وَكَفَنْ وِتْرُ كُسُوفُ عِيدُ اسْتِسقا سُنَنْ

143. فَجْرٌ رَغِيبَةٌ وَتُقْضَى لِلزَّوَالْ وَالْفَرْضُ يُقْضَى أَبَداً وَبِالتَّوَالْ

144. نُدِبَ نَفْلٌ مُطْلَقاً وَأُكِّدَتْ تَحِيَّةٌ ضُحًى تَرَاوِيحُ تَلَتْ

145. وَقَبْلَ وِتْرٍ مِثْلَ ظُهْرٍ عَصْرِ وَبَعْدَ مَغْرِبٍ وَبَعْدَ ظُهْرِ

146. لِنَقْصِ سُنَّةٍ سَهْواً يُسَنّ قَبْلَ السَّلامِ سَجْدَتَانِ أَوْ سُنَنْ

147. إِنْ أُكِّدَتْ وَمَنْ يَزِدْ سَهْواً سَجَدْ بَعْدُ كَذَا وَالنَّقْصَ غَلِّبْ إِنْ وَرَدْ

148. وَاسْتَدْرِكِ الْقَبْلِيَّ مَعْ قُرْبِ السَّلاَمْ وَاسْتَدْرِكِ الْبَعْدِيَّ وَلَوْ مِنْ بَعْدِ عَامْ

149. عَنْ مُقْتَدٍ يَحْمِلُ هَذَيْنِ الإِمَامْ وَبَطَلَتْ بِعَمْدِ نَفْخٍ أَوْ كَلاَمْ

150. لِغَيْرِ إِصْلاَحٍ وَبِالْمُشْغِلِ عَنْ فَرْضٍ وَفِي الْوَقْتِ أَعِدْ إِذَا يُسَنّ

151. وَحَدَثٍ وَسَهْوِ زَيْدِ الْمِثْلِ قَهْقَهَةٍ وَعَمْدِ شُرْبٍ أَكْلِ

152. وَسَجْدَةٍ قَيْءٍ وَذِكْرِ فَرْضِ أَقَلَّ مِنْ سِتٍّ كَذِكْرِ الْبَعْضِ

153. وَفَوْتِ قَبْلِيٍّ ثَلاَثَ سُنَنِ بِفَصْلِ مَسْجِدٍ كَطُولِ الزَّمَنِ

154. وَاسْتَدْرِكِ الرُّكْنَ فَإِنْ حَالَ الرُّكُوعْ فَأَلْغِ ذَاتَ السَّهْوِ وَالْبِنَا يَطُوعْ

155. كَفِعْلِ مَنْ سَلَّمَ لَكِنْ يُحْرِمُ لِلْبَاقِي وَالطُّولُ الْفَسَادَ مُلْزِمُ

156. مَنْ شَكَّ فِي رُكْنٍ بَنى عَلَى الْيَقِينْ وَلْيَسْجُدُوا الْبَعْدِيَّ لَكِنْ قَدْ يَبِينْ

157. لِأَنْ بَنَوْا فِي فِعْلِهِمْ وَالْقَوْلِي نَقْصٌ بِفَوْتِ سُورَةٍ فَالْقَبْلِي

158. كَذَاكِرِ الْوُسْطَى وَالْأَيْدِي قَدْ رَفَعْ وَرُكَباً لَا قَبْلَ ذَا لَكِنْ رَجَعْ

صَلَاةُ الْجُمُعَةِ

159. **فَصْلٌ** بِمَوْطِنِ الْقُرَى قَدْ فُرِضَتْ صَلَاةُ جُمْعَةٍ لِخُطْبَةٍ تَلَتْ

160. بِجَامِعٍ عَلَى مُقِيمٍ مَا انْعَذَرْ حُرٍّ قَرِيبٍ بِكَفَرْسَخٍ ذَكَرْ

161. وَأَجْزَأَتْ غَيْراً نَعَمْ قَدْ تُنْدَبُ عِنْدَ النِّدَا السَّعْيُ إِلَيْهَا يَجِبُ

162. وَسُنَّ غَسْلٌ بِالرَّوَاحِ اتَّصَلَا وَنُدِبَ تَهْجِيرٌ وَحَالٌ جَمُلَا

163. بِجُمْعَةٍ جَمَاعَةٌ قَدْ وَجَبَتْ سُنَّتْ بِفَرْضٍ وَبِرَكْعَةٍ رَسَتْ

164. وَنُدِبَتْ إِعَادَةُ الْفَذِّ بِهَا لَا مَغْرِباً كَذَا عِشًا مُوتِرُهَا

شُرُوطُ الْإِمَامِ

165. شَرْطُ الْإِمَامِ ذَكَرٌ مُكَلَّفُ آتٍ بِالْأَرْكَانِ وَحُكْمًا يَعْرِفُ

166. وَغَيْرُ ذِي فِسْقٍ وَلَحْنٍ وَاقْتِدَا فِي جُمْعَةٍ حُرٌّ مُقِيمٌ عُدِّدَا

167. وَيُكْرَهُ السَّلَسُ وَالْقُرُوحُ مَعْ بَادٍ لِغَيْرِهِمْ وَمَنْ يُكْرَهُ دَعْ

168. وَكَالْأَشَلِّ وَإِمَامَةٍ بِلَا رِداً بِمَسْجِدٍ صَلَاةٌ تُجْتَلَى

169. بَيْنَ الْأَسَاطِينِ وَقُدَّامَ الْإِمَامْ جَمَاعَةٌ بَعْدَ صَلَاةِ ذِي الْتِزَامْ

170. وَرَاتِبٌ مَجْهُولٌ أَوْ مَنْ أُبِنَا وَأَغْلَفٌ عَبْدٌ خَصِيٌّ ابْنُ زِنَا

171. وَجَازَ عِنِّينٌ وَأَعْمَى أَلْكَنُ — مُجَذَّمٌ خَفَّ وَهَذَا الْمُمْكِنُ

172. وَالْمُقْتَدِي الْإِمَامَ يَتْبَعُ خَلَا — زِيَادَةً قَدْ حُقِّقَتْ عَنْهَا لَا

173. وَأَحْرَمَ الْمَسْبُوقُ فَوْراً وَدَخَلْ — مَعَ الْإِمَامِ كَـــيْفَمَا كَانَ الْعَمَلْ

174. مُكَبِّراً إِنْ سَاجِداً أَوْ رَاكِعَا — أَلْفَاهُ لَا فِي جَلْسَةٍ وَتَابَعَا

175. إِنْ سلَّمَ الْإِمَامُ قَامَ قاضِيَا — أَقْوَالَهُ وَفِي الْأَفْعَالِ بَانِيَا

176. كَبَّرَ إِنْ حَصَّلَ شَفْعاً أَوْ أَقَلّْ — مِنْ رَكْعَةٍ وَالسَّهْوَ إِذْ ذَاكَ احْتَمَلْ

177. وَيَسْجُدُ الْمَسْبُوقُ قَبْلِيَّ الْإِمَامْ — مَعْهُ وَبَعْدِيّاً قَضَى بَعْدَ السَّلَامْ

178. أَدْرَكَ ذَاكَ السَّهْوَ أَوْ لَا قَيَّدُوا — مَنْ لَمْ يُحَصِّلْ رَكْعَةً لَا يَسْجُدُ

179. وَبَطَلَتْ لُمقْتَدٍ بِمُبْطِلِ — عَلَى الْإِمَامِ غَيْرَ فَرْعٍ مُنْجَلِي

180. مَنْ ذَكَرَ الْحَدَثَ أَوْ بِهِ غُلِبْ — إِنْ بَادَرَ الْخُرُوجَ مِنْهَا وَنُدِبْ

181. تَقْدِيمُ مُؤْتَمٍّ يُتِمُّ بِهِمُو — فَإِنْ أَبَاهُ انْفَرَدُوا أَوْ قَدَّمُوا

كِتَابُ الزَّكَاةِ

182. فُرِضَتِ الزَّكَاةُ فِيمَا يُرْتَسَمْ — عَيْنٍ وَحَبٍّ وَثِمَارٍ وَنَعَمْ

183. فِى الْعَيْنِ وَالْأَنْعَامِ حَقَّتْ كُلَّ عَامْ — يَكْمُلُ وَالْحَبُّ بِالْإِفْرَاكِ يُرَامْ

184. وَالتَّمْرُ وَالزَّبِيبُ بِالطِّيبِ وَفِي ذِي الزَّيْتِ مِنْ زَيْتِهِ وَالْحَبُّ يَفِي

185. وَهِيَ فِي الثِّمَارِ وَالْحَبِّ الْعُشُرْ أَو نِصْفُهُ إِنْ آلَةَ السَّقْيِ يَجُرّْ

186. خَمْسَةُ أَوْسُقٍ نِصَابٌ فِيهِمَا فِي فِضَّةٍ قُلْ مِائَتَانِ دِرْهَمَا

187. عِشْرُونَ دِينَاراً نِصَابٌ فِي الذَّهَبْ وَرُبُعُ الْعُشُرِ فِيهِمَا وَجَبْ

188. وَالْعَرْضُ ذُو التَّجْرِ وَدَيْنُ مَنْ أَدَارْ قِيمَتُـهَا كَالْعَيْنِ ثُمَّ ذُو احْتِكَارْ

189. زَكَّى لِقَبْضِ ثَمَنٍ أَو دَيْنِ عَيْناً بِشَرْطِ الْحَوْلِ لِلأَصْلَيْنِ

190. فِي كُلِّ خَمْسَةٍ جِمَالٌ جَذَعَةْ مِنْ غَنَمٍ بِنْتُ الْمَخَاضِ مُقْنِعَةْ

191. فِي الْخَمْسِ وَالْعِشْرِينَ وَابْنَةُ اللَّبُونْ فِي سِتَّةٍ مَعَ الثَّلاَثِينَ تَكُونْ

192. سِتّاً وَأَرْبَعِينَ حِقَّةٌ كَفَتْ جَذْعَةٌ إِحْدَى وَسِتِّينَ وَفَتْ

193. بِنْتَا لَـبُونٍ سِتَّةً وَسَبْعِينْ وَحِقَّتَانِ وَاحِداً وَتِسْعِينْ

194. وَمَعْ ثَلاَثِينَ ثَلاَثٌ أَيْ بَنَاتْ لَبُونٍ أَوْ خُذْ حِقَّتَيْنِ بِافْتِـيَاتْ

195. إِذَا الثَّلاَثِينَ تَلَتْـهَا الْمِائَةُ فِي كُلِّ خَمْسِينَ كَمَالاً حِقَّةُ

196. وَكُلُّ أَرْبَعِينَ بِنْتٌ لِلَّـبُونْ وَهَكَذَا مَا زَادَ أَمْرُهُ يَهُونْ

197. عِجْلٌ تَبِيعٌ فِي ثَلاَثِينَ بَقَرْ مُسِنَّةٌ فِي أَرْبَعِينَ تُسْتَطَرْ

198. وَهَكَذَا مَا ارْتَفَعَتْ ثُمَّ الْغَنَمْ شَاةٌ لِأَرْبَعِينَ مَعْ أُخْرَى تُضَمّ

199. فِي وَاحِدٍ عِشْرِينَ يَتْلُو وَمِئَةْ وَمَعْ ثَمَانِينَ ثَلَاثٌ مُجْزِئَةْ

200. وَأَرْبَعاً خُذْ مِنْ مِئِينَ أَرْبَعِ شَاةٌ لِكُلِّ مِائَةٍ إِنْ تُرْفَعِ

201. وَحَوْلُ الْأَرْبَاحِ وَنَسْلٍ كَالْأُصُولْ وَالطَّارِ لَا عَمَّا يُزَكَّى أَنْ يَحُولْ

202. وَلَا يُزَكَّى وَقَصٌ مِنَ النَّعَمْ كَذَاكَ مَا دُونَ النِّصَابِ وَلْيَعُمّ

203. وَعَسَلٌ فَاكِهَةٌ مَعَ الْخُضَرْ إِذْ هِيَ فِي الْمُقْتَاتِ مِمَّا يُدَّخَرْ

204. وَيَحْصُلُ النِّصَابُ مِنْ صِنْفَيْنِ كَذَهَبٍ وَفِضَّةٍ مِنْ عَيْنِ

205. وَالضَّأْنُ لِلْمَعْزِ وَبُخْتٌ لِلْعِرَابْ وَبَقَرٌ إِلَى الْجَوَامِيسِ اصْطِحَابْ

206. وَالْقَمْحُ لِلشَّعِيرِ لِلسُّلْتِ يُصَارْ كَذَا الْقَطَانِي وَالزَّبِيبُ وَالثِّمَارْ

207. مَصْرِفُهَا الْفَقِيرُ وَالْمِسْكِينُ غَازٍ وَعِتْقٌ عَامِلٌ مَدِينُ

208. مُؤَلَّفُ الْقَلْبِ وَمُحْتَاجٌ غَرِيبْ أَحْرَارُ إِسْلَامٍ وَلَمْ يُقْبَلْ مُرِيبْ

زَكَاةُ الْفِطْرِ

209. زَكَاةُ الْفِطْرِ صَاعٌ وَتَجِبْ عَنْ مُسْلِمٍ وَمَنْ بِرِزْقِهِ طُلِبْ

210. مِنْ مُسْلِمٍ بِجُلِّ عَيْشِ الْقَوْمِ لِتُغْنِ حُرّاً مُسْلِماً فِي الْيَوْمِ

كِتَابُ الصِّيَامِ

211. صِيَامُ شَهْرِ رَمَضَانَ وَجَبَا | فِي رَجَبٍ شَعْبَانَ صَوْمٌ نُدِبَا

212. كَتِسْعِ حِجَّةٍ وَأَحْرَى الآخِرُ | كَذَا الْمُحَرَّمُ وَأَحْرَى الْعَاشِرُ

213. وَيَثْبُتُ الشَّهْرُ بِرُؤْيَةِ الْهِلاَلْ | أَوْ بِثَلاَثِينَ قُبَيْلاً فِي كَمَالْ

214. فَرْضُ الصِّيَامِ نِيَّةٌ بِلَيْلِهِ | وَتَرْكُ وَطْءٍ شُرْبِهِ وَأَكْلِهِ

215. وَالْقَيْءِ مَعْ إِيصَالِ شَيْءٍ لِلْمِعَدْ | مِنْ أُذُنٍ أَوْ عَيْنٍ أَوْ أَنْفٍ وَرَدْ

216. وَقْتَ طُلُوعِ فَجْرِهِ إِلَى الْغُرُوبْ | وَالْعَقْلُ فِي أَوَّلِهِ شَرْطُ الْوُجُوبْ

217. وَلْيَقْضِ فَاقِدُهُ وَالْحَيْضُ مَنَعْ | صَوْماً وَتَقْضِي الْفَرْضَ إِنْ بِهِ ارْتَفَعْ

218. وَيُكْرَهُ اللَّمْسُ وَفِكْرٌ سَلِمَا | دَأْباً مِنَ الْمَذْيِ وَإِلاَّ حَرُمَا

219. وَكَرِهُوا ذَوْقَ كَقِدْرٍ وَهَذَرْ | غَالِبُ قَيْءٍ وَذُبَابٍ مُغْتَفَرْ

220. غُبَارُ صَانِعٍ وَطُرْقٍ وَسِوَاكْ | يَابِسٍ إِصْبَاحُ جَنَابَةٍ كَذَاكْ

221. وَنِـيَّةٌ تَـكْفِي لِمَا تَتَابُعُهْ | يَجِبُ إِلَّا إِنْ نَفَاهُ مَانِعُهْ

222. نُدِبَ تَعْجِيلٌ لِفِطْرٍ رَفَعَهْ | كَذَاكَ تَأْخِيرُ سُحُورٍ تَبِعَهْ

223. مَنْ أَفْطَرَ الْفَرْضَ قَضَاهُ وَلْيَزِدْ | كَفَّارَةً فِي رَمَضَانَ إِنْ عَمَدْ

224. لأَكْلِ أَوْ شُرْبِ فَمٍ أَوْ لِلْمَنِي   وَلَوْ بِفِكْرٍ أَوْ لِرَفْضِ مَا بُنِي

225. بِلاَ تَأَوُّلٍ قَرِيبٍ وَيُباحْ   لِضُرٍّ أَوْ سَفَرِ قَصْرٍ أَيْ مُبَاحْ

226. وَعَمْدُهُ فِي النَّفْلِ دُونَ ضُرٍّ   مُحَرَّمٌ وَلْيَقْضِ لاَ فِي الْغَيْرِ

227. وَكَفِّرَنْ بِصَوْمِ شَهْرَيْنِ وِلاَ   أَوْ عِتْقِ مَمْلُوكٍ بِالإِسْلاَمِ حَلاَ

228. وَفَضَّلُوا إِطْعَامَ سِتِّينَ فَقِيرْ   مُدًّا لِمِسْكِينٍ مِنَ الْعَيْشِ الْكَثِيرْ

كِتَابُ الْحَجِّ

229. اَلْحجُّ فَرْضٌ مَرَّةً فِي الْعُمْرِ   أَرْكَانُهُ إِنْ تُرِكَتْ لَمْ تُجْبَرِ

230. اَلْإِحْرَامُ وَالسَّعْيُ وُقُوفُ عَرَفَةْ   لَيْلَةَ الْأَضْحَى وَالطَّوَافُ رَدِفَهْ

231. وَالْوَاجِبَاتُ غَيْرُ الْأَرْكَانِ بِدَمْ   قَدْ جُبِرَتْ مِنْهَا طَوَافُ مَنْ قَدِمْ

232. وَوَصْلُهُ بِالسَّعْيِ مَشْيٌ فِيهِمَا   وَرَكْعَةُ الطَّوَافِ إِنْ تَحَتَّمَا

233. نُزُولُ مُزْدَلِفٍ فِي رُجُوعِنَا   مَبِيتُ لَيْلاَتٍ ثَلاَثٍ بِمِنَى

234. إِحْرَامُ مِيقَاتٍ فَذُو الحُلَيْفَهْ   لِطَيْبَ لِلشَّامِ وَمِصْرَ الْجُحْفَةْ

235. قَرْنٌ لِنَجْدٍ ذَاتُ عِرْقٍ لِلْعِرَاقْ   يَلَمْلَمُ الْيَمَنِ آتِيهَا وِفَاقْ

236. تَجَرُّدٌ مِنَ المَخِيطِ تَلْبِيَةْ   وَالحَلْقُ مَعْ رَمْيِ الْجِمارِ تَوْفِيَةْ

237. وَإِنْ تُرِدْ تَرْتِيبَ حَجِّكَ اسْمَعَا   بَيَانَهُ وَالذِّهْنَ مِنْكَ اسْتَجْمِعَا

238. إِنْ جِئْتَ رَابِغاً تَنَظَّفْ وَاْغتَسِلْ   كَـــوَاجِبٍ وَبِالشُّرُوعِ يَتَّصِلْ

239. وَالْبَسْ رِداً وَأُزْرَةً نَعْلَيْنِ   وَاسْتَصْحِبِ الْهَدْيَ وَرَكْــعَتَيْنِ

240. بِالْـــكَافِرُونَ ثُمَّ الْإِخْلاَصُ هُمَا   فَإِنْ رَكِـــبْتَ أَوْ مَشَيْتَ أَحْرِمَا

241. بِنِيَّةٍ تَصْحَبُ قَوْلاً أَوْ عَمَلْ   كَــمَشْيٍ أَوْ تَلْبِيَةٍ مِمَّا اتَّصَلْ

242. وَجَدِّدَنْهَا كُلَّمَا تَجَدَّدَتْ   حَالٌ وَإِنْ صَلَّيْتَ ثُمَّ إِنْ دَنَتْ

243. مَكَّةُ فَاغْتَسِلْ بِذِي طُوىً بِلاَ   دَلْكٍ وَمِنْ كُدَا الثَّنِيَّةِ ادْخُلاَ

244. إِذَا وَصَلْتَ لِلْبُيُوتِ فَاتْرُكَا   تَلْبِيَةً وَكُلَّ شُغْلٍ وَاسْلُكَا

245. لِلْبَيتِ مِنْ بَابِ السَّلاَمِ وَاسْتَلِمْ   الْحَجَرَ الْأَسْوَدَ كَــــبِّرْ وَأَتِمّ

246. سَبْعَةَ أَشْوَاطٍ بِهِ وقَدْ يَسَرْ   وَكَــــبِّرَنْ مُقَبِّلاً ذَاكَ الْحَجَرِ

247. مَتَى تُحاذِيهِ كَــذَا الْيَمَانِي   لَــكِنَّ ذَا بالْــيَدِ خُذْ بَــيَانِي

248. إِنْ لَمْ تَصِلْ لِلْحَجَرِ الْمَسْ بالْــيَدِ   وَضَعْ عَلَى الْفَمِ وكَــــبِّرْ تَقْتَدِ

249. وَارْمُلْ ثَلاَثاً وَامْشِ بَعْدُ أَرْبَعَا   خَلْفَ الْمَقَامِ رَكْــعَتَيْنِ أَوْقِعَا

250. وَادْعُ بِمَا شِئْتَ لَدَى الْمُلْتَزَمِ   وَالْحَجَرَ الْأَسْوَدَ بَعْدُ اسْتَلِمِ

251. وَاخْرُجْ إِلَى الصَّفَا وَقِفْ مُسْتَقْبِلاَ   عَلَيْهِ ثُمَّ كَــبِّرَنْ وَهَلِّلَا

252. وَاسْعَ لِـمَرْوَةٍ فَقِفْ مِثْلَ الصَّفَا   وَخُبَّ في بَطْنِ الْمَسِيلِ ذَا اقْتِفَا

253. أَرْبَعَ وَقْفَاتٍ بِكُلٍّ مِنْهُمَا   تَقِفُ وَالْأَشْوَاطَ سَبْعاً تَـمِّمَا

254. وَادْعُ بِماَ شِئْتَ بِسَعْيٍ وَطَوَافْ   وَبِالصَّفَا وَمَرْوَةٍ مَعَ اعْتِرَافْ

255. وَيَجِبُ الطُّهْرَانِ وَالسِّتْـرُ عَلَى   مَنْ طَافَ نَدْبُهَا بِسَعْيٍ يُجْتَلَى

256. وَعُدْ فَلَبِّ لِـمُصلَّى عَرَفَةْ   وَخُطْبَةُ السَّابِعِ تَأْتِي لِلصِّفَةْ

257. وَثامِنَ الشَّهْرِ اخْرُجَنَّ لِـمِنَى   بِعَرَفَاتٍ تَاسِعاً نُزُولُنَا

258. وَاغْتَسِلَنْ قُرْبَ الزَّوَالِ وَاحْضُرَا   الْخُطْـبَتَيْنِ وَاجْمَعَنْ وَقَصِّرَا

259. ظُهْرَيْكَ ثُمَّ الْجَبَلَ اصْعَدْ رَاكِبَا   عَلَى وُضُوءٍ ثُمَّ كُنْ مُوَاظِبَا

260. عَلَى الدُّعَا مُهَلِّلاً مُبْتَهِلَا   مُصَلِّيًا عَلَى النَّبِي مُسْتَقْبِلَا

261. هُنَيْهَةً بَعْدَ غُرُوبِهَا تَقِفْ   وَانْفِرْ لِمُزْدَلِفَةٍ وَتَنْصَرِفْ

262. فِي الْمَأْزِمَيْنِ الْعَلَمَيْنِ نَكِّبِ   وَقْصِرْ بِهَا وَاجْمَعْ عِشاً لِمَغْرِبِ

263. وَاحْطُطْ وَبِتْ بِهَا وَأَحْيِ لَيلَتَكْ   وَصَلِّ صُبْحَكَ وَغَلِّسْ رِحْلَتَكْ

264. قِفْ وَادْعُ بِالْمشْعَرِ لِلْأَسفارِ   وَأَسْرِعَنْ فِي بَطْنِ وَادِي النَّارِ

265. وسِرْ كَـمَا تَكُونُ لِلْعَقَبَةِ   فارْمِ لَدَيْهَا بِحِجَارٍ سَبْعَةِ

266. مِنْ أَسْفَلٍ تُسَاقُ مِنْ مُزْدَلِفَهْ   كَالْفُولِ وَانْحَرْ هَدْياً إِنْ بِعَرَفَهْ

267. أَوْقَفْتَهُ وَاحْلِقْ وَسِرْ لِلْبَيْتِ   فَطُفْ وَصَلِّ مِثْلَ ذَاكَ النَّعْتِ

268. وَارْجِعْ فَصَلِّ الظُّهْرَ فِي مِنًى وَبِتْ   إِثْرَ زَوَالِ غَدِهِ أَرْمِ لاَ تُفِتْ

269. ثَلَاثَ جَمْرَاتٍ بِسَبْعِ حَصَيَاتْ   لِكُلِّ جَمْرَةٍ وَقِفْ لِلدَّعَوَاتْ

270. طَوِيلاً إِثْرَ الْأَوَّلَيْنَ أَخِّرَا   عَقَبَةً وَكُلَّ رَمْيٍ كَـبِّرَا

271. وَافْعَلْ كَذَاكَ ثَالِثَ النَّحْرِ وَزِدْ   إِنْ شِئْتَ رَابِعاً وَتَمَّ مَا قُصِدْ

272. وَمَنَعَ الإِحْرَامُ صَيْدَ الْبَرِّ   فِي قَتْلِهِ الْجَزَاءُ لَا كَالْفَارِ

273. وَعَقْرَبٍ مَعَ الْحِدَا كَلْبٍ عَقُورْ   وَحَيَّةٍ مَعَ الْغُرَابِ إِذْ يَجُورْ

274. وَمَنَعَ الْمُخِيطَ بِالْعُضْوِ وَلَوْ   بِنَسْجٍ أَوْ عَقْدٍ كَخَاتَمٍ حَكَوْا

275. وَالسَّتْرُ لِلوَجْهِ أَوِ الرَّأْسِ بِمَا   يُعَدُّ سَاتِراً وَلَكِنْ إِنَّمَا

276. تُمْنَعُ الْأُنْـثَى لُـبْسَ قُفَّازٍ كَذَا   سَتْـرٌ لِوَجْهٍ لَا لِسَتْرٍ أُخِذَا

277. وَمَنَعَ الطِّيبَ وَدُهْناً وَضَرَرْ   قَمْلٍ وَإِلْقَا وَسَخٍ ظُفْرٍ شَعَرْ

278. وَيَفْتَدِي لِفِعْلِ بَعْضِ مَا ذُكِرْ مِنَ الْمُخِيطِ لِهُنَا وَإِنْ عُذِرْ

279. وَمَنَعَ النِّسَا وَأَفْسَدَ الْجِمَاعْ إِلَى الْإِفَاضَةِ يُبْقَى الْامْتِنَاعْ

280. كَالصَّيْدِ ثُمَّ بَاقِي مَا قَدْ مُنِعَا بِالْجَمْرَةِ الْأُولَى يَحِلُّ فَاسْمَعَا

281. وَجَازَ الاسْتِظْلاَلُ بِالْمُرْتَفِعِ لَا فِي الْمَحَامِلِ وَشُقْدُفٍ فَعِ

282. وَسُنَّةَ الْعُمْرَةِ فَافْعَلْهَا كَمَا حَجٍّ وَفِي التَّنْعِيمِ نَدْباً أَحْرِمَا

283. وَإِثْرَ سَعْيِكَ احْلِقَنْ وَقَصِّرَا تَحِلُّ مِنْهَا وَالطَّوَافَ كَثِّرَا

284. مَا دُمْتَ فِي مَكَّةَ وَارْعَ الْحُرْمَةْ لِجَانِبِ الْبَيْتِ وَزِدْ فِي الْخِدْمَةْ

285. وَلازِمِ الصَّفَّ فَإِنْ عَزَمْتَ عَلَى الْخُرُوجِ طُفْ كَمَا عَلِمْتَ

286. وَسِرْ لِقَبْرِ الْمُصْطَفَى بِأَدَبِ وَنِيَّةٍ تُجَبْ لِكُلِّ مَطْلَبِ

287. سَلِّمْ عَلَيْهِ ثُمَّ زِدْ لِلصِّدِّيقْ ثُمَّ إِلَى عُمَرَ نِلْتَ التَّوفِيقْ

288. وَاعْلَمْ بِأَنَّ ذَا الْمَقَامِ يُسْتَجَابْ فِيهِ الدُّعَا فَلَا تَمَلَّ مِنْ طِلاَبْ

289. وَسَلْ شَفَاعَةً وَخَتْماً حَسَنَا وَعَجِّلِ الْأَوْبَةَ إِذْ نِلْتَ الْمُنَا

290. وَادْخُلْ ضُحىً وَاصْحَبْ هَدِيَّةَ السُّرُورْ إِلَى الْأَقَارِبِ وَمَنْ بِكَ يَدُورْ

كِتَابُ مَبَادِئُ التَّصَوُّفِ وَهَوَادِي التَّعَرُّفِ

291. وَتَوْبَةٌ مِنْ كُلِّ ذَنْبٍ يُجتَرَمْ　　تَجِبُ فَوْراً مُطْلَقاً وَهِيَ النَّدَمْ

292. بِشَرْطِ الْإِقْلاَعِ وَنَفْيِ الإِصْرَارْ　　وَلْيَتلاَفَ مُمْكِناً ذَا اسْتِغْفَارْ

293. وحَاصِلُ التَّقْوَى اجْتِنَابٌ وَامْتِثَالْ　　فِي ظَاهِرٍ وَبَاطِنٍ بِذَا تُنَالْ

294. فَجَاءَتِ الأَقْسَامُ حَقّاً أَرْبَعَةْ　　وَهِيَ لِلسَّالِكِ سُبْلُ المَنْفَعَةْ

295. يَغُضُّ عَيْنَيْهِ عَنِ الْمَحَارِمْ　　يَكُفُّ سَمْعَهُ عَنِ الْمَآثِمْ

296. كَغِيبَةٍ نَمِيمَةٍ زُورٍ كَذِبْ　　لِسَانُهُ أَحْرَى بِتَرْكِ مَا جُلِبْ

297. يَحْفَظُ بَطْنَهُ مِنَ الْحَرَامِ　　يَتْرُكُ مَا شُبِّهَ بِاهْتِمَامِ

298. يَحْفَظُ فَرْجَهُ وَيَتَّقِي الشَّهِيدْ　　فِي الْبَطْشِ وَالسَّعْيَ لِمَمْنُوعٍ يُرِيدْ

299. وَيُوقِفُ الأُمُورَ حَتَّى يَعْلَمَا　　مَا اللَّهُ فِيهِنَّ بِهِ قَدْ حَكَمَا

300. يُطَهِّرُ الْقَلْبَ مِنَ الرِّيَاءِ　　وَحَسَدٍ عُجْبٍ وَكُلِّ دَاءِ

301. وَاعْلَمْ بِأَنَّ أَصْلَ ذِي الْآفَاتِ　　حُبُّ الرِّيَاسَةِ وَطَرْحُ الْآتِي

302. رَأْسُ الخَطَايَا هُوَ حُبُّ الْعَاجِلَةْ　　لَيْسَ الدَّوَا إِلاَّ فِي الاِضْطِرَارِ لَهْ

303. يَصْحَبُ شَيْخاً عَارِفَ الْمَسَالِكْ　　يَقِيهِ فِي طَرِيقِهِ الْمَهَالِكْ

304. يُذَكِّرُهُ اللَّهَ إِذَا رَآهُ وَيُوصِلُ الْعَبْدَ إِلَى مَوْلَاهُ

305. يُحَاسِبُ النَّفْسَ عَلَى الْأَنْفَاسِ وَيَزِنُ الْخَاطِرَ بِالْقِسْطَاسِ

306. وَيَحْفَظُ الْمَفْرُوضَ رَأْسَ المَالِ وَالنَّفْلَ رِبْحَهُ بِهِ يُوَالِي

307. وَيُكْثِرُ الذِّكْرَ بِصَفْوِ لُبِّهِ وَالْعَوْنُ فِي جَمِيعِ ذَا بِرَبِّهِ

308. يُجَاهِدُ النَّفْسَ لِرَبِّ الْعَالَمِينْ وَيَتَحلَّى بِمَقَامَاتِ الْيَقِينْ

309. خَوْفٌ رَجاً شُكْرٌ وَصَبْرٌ تَوْبَةْ زُهْدٌ تَوَكُّلٌ رِضاً مَحَبَّةْ

310. يَصْدُقُ شَاهِدَهُ فِي الْـمُعَامَلَةْ يَرْضَى بِمَا قَدَّرَهُ الْإِلَهُ لَهْ

311. يَصِيرُ عِنْدَ ذَاكَ عَارِفاً بِهِ حُرّاً وَغَيْرُهُ خَلاَ مِنْ قَلْبِهِ

312. فَحَبَّهُ الْإِلَهُ وَاصْطَفَاهُ لِـحَضْرَةِ الْقُدُّوسِ وَاجْتَبَاهُ

313. ذَا الْقَدْرُ نَظْماً لَا يَفِي بِالْغَايَةْ وَفِي الَّذِي ذَكَرْتُهُ كِفَايَةْ

314. أَبْيَاتُهُ أَرْبَعَةٌ عَشَرَ تَصِلْ مَعَ ثَلَاثِمِائَةٍ عَدُّ الرُّسُلْ

315. سَمَّيْتُهُ بِالْمُرْشِدِ الْمُعِينِ عَلَى الضَّرُورِي مِنْ عُلُومِ الدِّينِ

316. فَأَسْأَلُ النَّفْعَ بِهِ عَلَى الدَّوَامْ مِنْ رَبِّنَا بِجَاهِ سَيِّدِ الْأَنَامْ

317. قَدِ انْتَهَى وَالحَمْدُ لِلَّهِ الْعَظِيمْ صَلَّى وَسَلَّمَ عَلَى الْهَادِي الْكَرِيمْ

Bibliography

COMMENTARIES OF THE QUR'AN

- *Hashiya as-Sawi 'ala Tafsir al-Jalalayn*, by Ahmad as-Sawi, Dar al-Fikr, Beirut
- *At-Tas-hil li 'Ulum at-Tanzil*, by Ibn Juzayy al-Kalbi, Dar al-Arqam, Beirut
- *Tafsir al-Khazin*, by 'Alauddin 'Ali Ibn Muhammad al-Baghdadi al-Khazin, Dar al-Kutub al-'Ilmiyya, Beirut
- *Ahkam al-Qur'an*, by Abu Bakr Ibn al-'Arabi, Dar al-Kutub al-'Ilmiyya, Beirut
- *Al-Jami' li Ahkam al-Qur'an*, by Abu 'Abdullah ibn Ahmad al-Ansari al-Qurtubi, Dar al-Hadith, Cairo

HADITH COLLECTIONS AND THEIR COMMENTARIES

- *Al-Muwatta'*, by al-Imam Malik Ibn Anas, Dar Ihya' al-Kutub al-'Arabiyya, Cairo
- *Sharh al-Muwatta'*, by az-Zurqani, Dar al-Fikr, Beirut
- *Al-Muntaqa Sharh al-Muwatta'*, by al-Baji, Dar al-Kutub al-'Ilmiyya, Beirut
- *Sahih* al-Imam al-Bukhari, Dar al-Basha'ir al-Islamiyya, Beirut
- *Fath al-Bari Sharh Sahih al-Bukhari*, by Ibn Hajar al-'Asqalani, Dar al-Ma'rifa, Beirut
- *Irshad as-Sari Sharh Sahih al-Bukhari*, by Shihab ad-Din al-Qastalani, Dar al-Fikr, Beirut
- *Sahih* al-Imam Muslim, Dar al-Jil, Beirut
- *Sharh Sahih Muslim*, by an-Nawawi, Dar al-Hadith, Cairo
- *Ikmal al-Mu'allim Bifawa'id Muslim*, by al-Qadi 'Iyad, Dar al-Kutub al-'Ilmiyya
- *Ikmal Ikmal al-Mu'allim*, by Muhammad Ibn Khalaf al-Washtani al-Ubbi, Dar al-Kutub al-'Ilmiyya, Beirut
- *Sunan* al-Imam at-Tirmidhi, Dar al-Kutub al-'Ilmiyya, Beirut

- *Tuhfat al-Ahwadhi Sharh Jami' at-Tirmidhi*, by Muhammad Ibn 'Abdirrahman al-Mubarakfuri, Dar al-Kutub al-'Ilmiyya, Beirut
- *Sunan* al-Imam Abi Dawud, Dar al-Kutub al-'Ilmiyya, Beirut
- *'Awn al-Ma'bud, Sharh Sunan Abi Dawud*, by Muhammad Shams al-Haqq al-'Adhimabadi, Dar al-Kutub al-'Ilmiyya, Beirut *Sunan* al-Imam an-Nasa'i
- *Sunan* al-Imam Ibn Majah, Dar al-Kutub al-'Ilmiyya, Beirut
- *Musnad* al-Imam Ahmad, Dar al-Kutub al-'Ilmiyya, Beirut
- *Takhrij Ahadith al-Ihya'*, al-Hafidh al-'Iraqi, Dar al-Ma'rifa, Beirut
- *Al-Jami' as-Saghir*, by Jalaluddin as-Suyuti, Dar al-Kutub al-'Ilmiyya, Beirut
- *Fayd al-Qadir Sharh al-Jami' as-Saghir*, by al-Munawi, Dar al-Kutub al-'Ilmiyya, Beirut
- *Ziyadat al-Jami' as-Saghir*, by Jalaluddin as-Suyuti, Mustafa al-Babi al-Halabi, Cairo
- *Nayl al-Awtar*, by ash-Shawkani, Dar al-Ma'rifa, Beirut
- *Sharh al-Arba'in an-Nawawiyyah*, by Ibn Daqiq al-'Id, Dar al-Fikr, Beirut

BOOKS OF 'AQIDA

- *Ra'ihat al-Janna, Sharh Ida'a ad-Dujunna fi 'Aqa'id Ahl as-*Sunna by 'Abdulghani an-Nablusi, Dar al-Kutub al-'Ilmiyya, Beirut
- *Tuhfat al-Murid Sharh Jawharat at-Tawhid*, by Ahmad al-Bayjuri ash-Shafi'i, Dar al-Kutub al-'Ilmiyyah, Beirut
- *Sharh Jawharat at-Tawhid*, by Bakri Rajab, Dar al-Khayr, Beirut
- *Ar-Ra'id fi 'Ilm al-'Aqa'id*, al-'Arabi al-Luh, Tetouan
- *Sharh al-'Aqa'id an-Nasafiyyah*, by Sa'd ad-Din at-Taftazani, Maktaba al-Kulliyat al-Azhariyya, Cairo
- *Al-Matalib al-Wafiyyah Sharh al-'Aqa'id an-Nasafiyyah*, by 'Abdullah al-Harawi, Dar al-Mashari'
- *Kubra al-Yaqiniyat al-Kawniyya*, by Muhammad Sa'id Ramadan al-Buti, Dar al-Fikr, Damascus
- *Tali' al-Bashra 'ala al-'Aqidat as-Sanusiyya as-Sugra*, by Ibrahim al-Maraghni, Tunis
- *Sharh Umm al-Barahin*, by Ahmad al-Ansari, Kano

BOOKS OF FIQH

- *Al-Habl al-Matin 'ala Nazm al-Murshid al-Mu'in*, by al-Marrakushi, Dar ar-Rashad al-Haditha, Casablanca
- *Mukhtasar ad-Durr ath-Thamin wa al-Mawrid al-Ma'in*, by Muhammad al-Fasi Mayyara, Maktaba al-Alma'arif, Beirut
- *Hashiya 'ala Sharh Mayyara as-Saghir 'ala al-Murshid al-Mu'in,* by at-Talib ibn al-Hajj, Dar ar-Rashad al-Haditha, Casablanca
- *Ar-Risala al-Fiqhiyyah* by Ibn Abi Zaid al-Qayrawani, Ministry of Awaqf and Islamic Affairs, Rabat
- *Ath-Thamar ad-Dani,* by Salih 'Abdusssami' al-Abi al-Azhari, Dar al-Kutub al-'Ilmiyya, Beirut
- *Al-Fawakih ad-Dawani 'ala Risala al-Qayrawani*, by an-Nafrawi, Dar al-Fikr, Beirut
- *Taqrib al-Ma'ani Sharh Risala al-Qayrawani*, by 'Abdul-majid ash-Sharnubi al-Azhari, Dar al-Kutub al-'Ilmiyya, Beirut
- *Hashiya al-'Adawi 'ala Kifayat at-Talib ar-Rabbani*, by al-'Adawi, Dar al-Fikr, Beirut
- *Masalik ad-Dilala fi Takhrij Ahadith ar-Risala*, by Ahmad Ibn Siddiq al-Ghumari, Dar al-Fikr, Beirut
- *Ash-Sharh as-Saghir 'ala Aqrab al-Masalik*, by ad-Dardir, Dar al-Ma'arif, Cairo.
- *Tabyin al-Masalik Sharh Tadrib as-Salik ila Aqrab al-Masalik*, by 'Abd al-'Aziz Ahmad al Mubarak al-Isa'i and Muhammad ash-Shaybani ash-Shinqiti, Dar al-Garb al-Islami, Beirut
- *Hashiya ad-Dasuqi 'ala ash-Sharh al-Kabir*, by Ahmad ibn 'Arafa ad-Dasuqi, Dar al-Kutub al-'Ilmiyya, Beirut
- *Hashiya al-Khurashi 'ala Mukhtasar Khalil*, by 'Abdullah ibn 'Ali al-Khurashi, Dar al-Kutub al-'Ilmiyya, Beirut
- *Minah al-Jalil Sharh Mukhtasar Khalil*, by Muhammad 'Illish, Dar al-Fikr, Beirut
- *Mawahib al-Jalil Sharh Mukhtasar Khalil*, by al-Hattab, Dar al-Kutub al-'Ilmiyya, Beirut
- *Al-Mudawwana al-Kubra*, Sahnun ibn Sa'id at-Tannukhi, Mu'assasa ar-Rayyan, Beirut
- *Al-Qawanin al-Fiqhiyyah*, by Ibn Juzayy al-Kalbi, Dar ar-Rashad al-Haditha, Casablanca
- *Bidayat al-Mujtahid wa Nihayat al-Muqtasid*, by Ibn Rushd al-

Hafid, Dar al-Ma'rifa, Beirut

- *Mudawwanat al-Fiqh al-Maliki wa Adillatuh*, 'Abdur-rahman al-Ghiryani, Mu'assasa ar-Rayyan, Beirut
- *Al-Mi'yar al-Mu'rib wa al-Jami' al-Mughrib 'an Fatawi Ahl Ifriqiyya wa al-Andalus wa al-Maghrib*, by al-Wansharisi, Ministry of Awaqf and Islamic Affairs, Rabat
- *Al-Fath al-'Ali al-Malik fi Fatawi 'ala Madhhab Malik*, by 'Illish, Dar al-Fikr, Beirut
- *Al-Fatawi al-Kubra*, by al-Mahdi al-Wazzani, Ministry of Awaqf and Islamic Affairs, Rabat
- *Al-Fatawi as-Sughra*, by al-Mahdi al-Wazzani, Ministry of Awaqf and Islamic Affairs, Rabat
- *Al-Fatwa wa al-Qada' fi al-Madhhab al-Maliki*, 'Ali al-'Iraqi, unpublished
- *Al-Mawsu'a al-Fiqhiyya al-Kuwaitiyya*, Ministry of Awqaf and Islamic Affairs of Kuwait
- *Al-Fiqh al-Islami wa Adillatuh*, by Wahbah az-Zuhayli, Dar al-Fikr, Damascus
- *Al-Fiqh 'ala al-Madhahib al-Arba'a*, by Abdel-Rahman al-Jaziri, Hakikat Kitabevi, Istanbul
- *Sina'a al-Fatwa wa fiqh al-Aqalliyat*, 'Abdur-rahman ibn Bayyah, Dar al-Minhaj, Jeddah
- *Al-Fiqh al-Maliki wa Adillatuh*, by al-Habib ibn Tahir, Mu'assasa al-Ma'arif

BOOKS OF USUL AL-FIQH

- *Nathr al-Wurud 'ala Maraqi as-Su'ud*, by Muhammad al-Amin ash-Shinqiti, Dar al-Manara, Jeddah
- *Hashiya al-Bannani 'ala Sharh Jalal ad-Din al-Mahalli 'ala Jam' al-Jawami'*, Taj ad-Din as-Subki, Dar al-Fikr, Beirut
- *Usul al-Fiqh al-Islami*, by Wahbah az-Zuhayli, Dar al-Fikr, Damascus

BOOKS OF TASAWWUF

- *Ar-Risala al-Qushayriyya*, by al-Qushayri, Dar al-Jil, Beirut
- *Sharh al-Hikam al-'Ata'iyya*, by ash-Sharnubi, Dar Ibn Kathir, Beirut

- *Iqadh al-Himam fi Sharh al-Hikam*, by Ibn 'Ajiba, al-Maktaba ath-Thaqafiyya, Beirut
- *Al-Hikam al-'Ata'iyya Sharh wa Tahlil*, by Sa'id Ramadan al-Buti, Dar al-Fikr, Damascus
- *Al-Futuhat al-Ilahiyya fi Sharh al-Mabahith al-Asliyya*, by Ibn Ajiba, Dar al-Kutub al-'Ilmiyya, Beirut
- *Ihya' 'Ulum ad-Din*, by Abu Hamid al-Ghazali, Dar al-Ma'rifa, Beirut
- *Qawa'id at-Tasawwuf*, by Ahmad Zarruq, Dar al-Jil, Beirut
- *Haqa'iq 'an at-Tasawwuf*, by 'Abdulqadir 'Isa, Diwan Press, Norwich
- *The Hundred Steps*, by Shaykh 'Abdulqadir as-Sufi, Madinah Press, Kuala Lumpur
- *Diwan Bughyat al-Murideen as-Sa'ireen wa Tuhfat as-Salikeen al-'Arifeen*, Shaykh Muhammad ibn al-Habib al-Amghari, Dar Sadir, Beirut
- *Al-I'lam bi anna at-Tasawwuf min ash-Shari'a al-Islam*, by al-Hafizh 'Abdullah al-Ghumari, Dar at-Tiba'a al-Haditha, Casablanca
- *Husn at-Talattuf fi Bayan Wujub Suluk at-Tasawwuf*, by al-Hafiz 'Abdullah al-Ghumari, Dar at-Tiba'a al-Haditha, Casablanca
- *Shifa' as-Sa'il fi Tahdhib al-Masa'il*, by Ibn Khaldun, Dar al-Fikr, Damascus

About the Author

HE IS ALI ibn Abdulhaq Al-'Iraqi al-Hussaini. A descendent of the Prophet (May Allah grant peace and blessings to him and his family) from his grandson Imam Hussain (May Allah be pleased with him). Born in Málaga (Spain) in 1965 and brought up in Tangier (Morocco).

He holds a degree (*Ijaza*) in Shari'a and Usul ad-Din by the European Faculty of Islamic Sciences (France). He has studied Islamic Sciences in the traditional way with various 'ulama like Shaykh Muhammad Al Kassbi (Morocco), Sidi 'Abd as-Salam Afkir (Morocco), Shaykh Muhammad an-Naifar (Tunisia), Shaykh Muhammad al-Lakhwa (Tunisia), Sidi Muhammad al-Wazzani (Morocco), holding *ijazat* from some of them.

He has translated into Spanish and commented on the *Risala* of Ibn Abi Zaid al-Qayrawani, the *Hikam* of Ibn 'Ata' Allah, the *Ajurrumiya* of Ibn Ajurrum, the *Dala'il al-Khairat* of Al-Jazuli, the *Burda* of Al-Busayri. He is also the author of *'Aqida – Two treatises on Muslim Creed*, a commentary on Surat al-Fatiha, an *Introduction to the Science of Tasawwuf*, a *Treatise on Islamic Governance* and a number of other works.

Under the guidance of Shaykh Abdalqadir as-Sufi, he has lived and taught in South Africa, Spain, the USA and the United Kingdom. He has been Imam Khatib in Newark NJ (USA), Madrid (Spain), the Jami'a Mosque of Cape Town (South Africa) and lecturer in Islamic Studies there at the Dallas College.